Praise for Jo Marchant's
Decoding the Heavens

"Marchant's account is engagingly human. . . . Captures both the endless elusiveness of the past and our ever-evolving attempts to understand it."

—*American Scientist*

"[Marchant's] gripping and varied account will propel the mechanism to greater fame."

—*Nature*

"A fascinating tale."

—*New Scientist*

"[*Decoding the Heavens*] takes readers on the device's fascinating journey from the bottom of the sea to cutting edge X-ray technology."

—*History Magazine*

"Science readers will be entranced by Marchant's vibrant depiction of the characters in this remarkable story of ancient technology."

—*Booklist*

"An enthralling tale of research and reconstruction. . . . Archaeology, conceptual thinking, mathematics, and the successful reconstruction of the mechanism all are engrossingly presented."

—*ForeWord*

"A fascinating new book . . . An intriguing journey into science, archaeology, and history."

—*Huffington Post*

"Jo Marchant . . . narrates the century-long quest to understand the purpose of the mechanism as if it were a dynastic saga, complete with jealous rivals, personal tragedies and lifelong obsessions."

—*Focus* (BBC) magazine

"Though it is more than 2,000 years old, the Antikythera Mechanism represents a level that our technology did not match until the 18th century, and must therefore rank as one of the greatest basic mechanical inventions of all time."

—Arthur C. Clarke

Decoding the *Heavens*

A 2,000-Year-Old Computer—
And the Century-Long Search
to Discover Its Secrets

Jo Marchant

DA CAPO PRESS
A MEMBER OF THE PERSEUS BOOKS GROUP

Printed in the United States of America. For information, address Da Capo Press, 11 Cambridge Center, Cambridge, MA 02142.

Designed by Palimpsest Book Production Limited, Grangemouth, Stirlingshire
Set in 11 point Bembo by the Perseus Books Group

Map by ML Design, London

Cataloging-in-Publication data for this book is available from the Library of Congress.

First Da Capo Press edition 2009
First Da Capo Press paperback edition 2010
Reprinted by arrangement with William Heinemann
HC ISBN: 978-0-306-81742-7
PB ISBN: 978-0-306-81861-5
Library of Congress Control Number: 2008939733

Published by Da Capo Press
A Member of the Perseus Books Group
www.dacapopress.com

10 9 8 7 6 5 4 3 2 1

To Ian

Contents

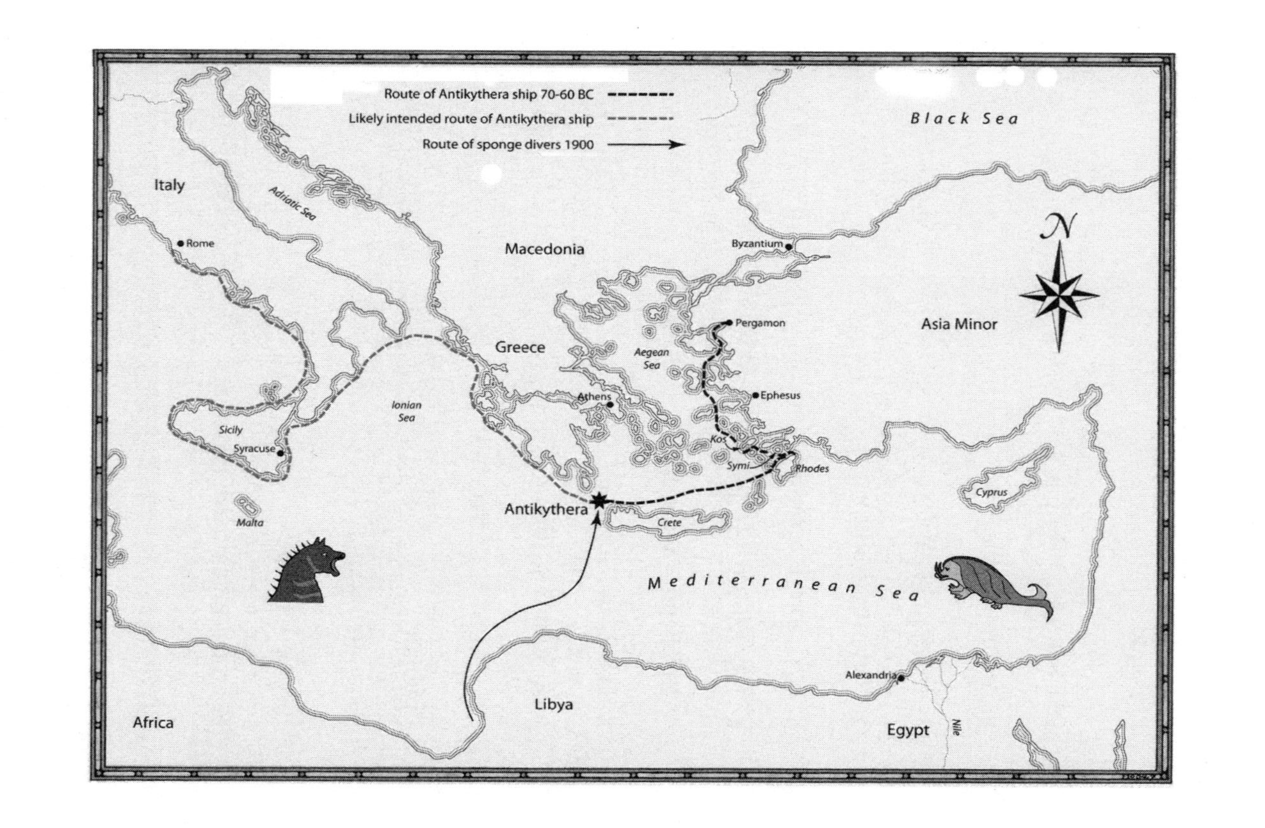
Route of Antikythera ship 70-60 BC
Likely intended route of Antikythera ship
Route of sponge divers 1900
Black Sea
Italy
Adriatic Sea
Rome
Macedonia
Byzantium
N
Pergamon
Asia Minor
Greece
Aegean Sea
Ionian Sea
Athens
Ephesus
Sicily
Syracuse
Kos
Symi
Rhodes
Cyprus
Antikythera
Crete
Malta
Mediterranean Sea
Alexandria
Libya
Africa
Egypt
Nile

Prologue

IN A CORNER of the National Archaeological Museum in Athens is something that doesn't fit. It is nothing like the classical Greek statues and vases that fill the rest of the echoing hall. Three flat pieces of what looks like mouldy, green cardboard are delicately suspended inside a glass case.

Within each piece, layers of what was once metal have been squashed together and are now covered with corrosion products – from the whitish green of tin oxide to the dark bluish green of copper chloride. They've been under the sea for 2,000 years, and it shows.

Look closer though, and you'll see something impossible. Through the deposits, shadowy outlines are visible: engraved letters, a large wheel and part of an encrusted but precisely marked circular scale. Next to these strange items an X-ray image shows what's hidden inside. Beneath the ancient, calcified surfaces, delicate cogwheels of all sizes are jostling for space, their triangular teeth so perfectly formed it seems that any second they might start clicking round. The design of the mechanism is modern and immediately recognisable. It looks just like the inside of an alarm clock.

This is the Antikythera mechanism. Its fragments are now

known to contain at least 30 gear wheels and urgent inscriptions are crammed onto every surviving surface. Rescued from an ancient shipwreck in 1901, it is one of the most stunning artefacts we have from antiquity and, according to everything we know about the technology of the time, it shouldn't exist. Nothing close to its sophistication appears again for well over a millennium, with the development of elaborate astronomical clocks in Renaissance Europe.

Never mind the statues that fill the rest of the museum. Never mind the riches from all the ancient shipwrecks discovered since. Beautiful and inspiring as they are, each individual piece of art merely fleshes out our appreciation of the Greek sculptor's craft. This unassuming object is different. Although 2,000 years under the sea have left it dull and battered, the ideas and expertise it embodies have turned upside down our understanding of who the ancient Greeks were and what they were capable of, igniting a mystery that has taken more than a century to decode.

So what was it? Who on Earth could have made it? And once this complex technology arose, what caused it to be forgotten for so long? Since 1901 a number of men have devoted their lives to solving the mechanism and answering these questions, each unable to turn away from the mystery once it had found them. Many of them didn't live to learn the whole truth, but each gleaned a part of it, and this book aims to tell their stories.

None of this could have happened, however, without Captain Kontos and his hardy crew of sponge divers, for without them the Antikythera fragments would still be languishing at the bottom of the sea. They discovered the wreck and risked their lives in the first ever attempt to salvage artefacts from a sunken ship; a daring adventure from which they did not all return.

I

I See Dead People

I might have reached my own land unscathed; but no, as I was doubling Cape Malea I was caught by wave and current and wind from the North and was driven off course and past Kythera. Then for nine days I was carried by ruthless winds over teeming ocean. On the tenth day we reached the land of the Lotus-Eaters, whose only fare is that fragrant fruit.

— HOMER, *THE ODYSSEY*

FOR THE ANCIENT Greeks, the ocean was the centre of the world. There was no single country with borders we'd recognise today as 'Greece'; instead the Greeks, bound by a common culture and language, retained their identity as they spread far across the Mediterranean Sea. By Homer's time, around the eighth century BC, Greek speakers from the ancient provinces of Attica, Boeotia, Laconia and Achaea had reached many far off lands – Macedonia and Thrace in the north; the scattered islands of the Aegean as well as Anatolia and the Asia Minor coast in the east; Egypt and Libya to the south; and Italy, Sicily, and France to the west.

The only practical way to get between these far-flung settlements was by water. For thousands of years, ships – not just from Greece, but also the rival civilisations of Egypt, Phoenicia and later Rome – crisscrossed the Mediterranean. As well as settlers, they carried soldiers, slaves, diplomats and merchants. Goods transported as gifts and for trade included staples such as grain, wine and olive oil, but there were luxuries too from every corner: ostrich eggs from Libya, gold and ivory from Egypt, lapis lazuli from Afghanistan. Merchants carried amber beads from northern Europe and from the mines of Cyprus they brought copper – to forge the sought-after bronze weapons, armour and statues.

At the centre of this watery world lay the mountainous peninsula we now call Greece. To get between the island-filled Aegean Sea in the east and the more open waters of the west, captains like Odysseus had to navigate their ships through the treacherous and stormy passage between the peninsula's southern tip, Cape Malea, and the island of Crete.

Nearly 3,000 years after Homer's tale, this gateway hadn't lost any of its malice. More than a hundred generations since *The Odyssey* entranced its first listeners, another crew of Greek sailors was trying to pass Cape Malea, on the way home to the Aegean island of Symi. But they, too, were blown off course and taken on an epic adventure of their own.

It was the year 1900. The world was now dominated by the expanding British Empire of Queen Victoria and the spreading iron fingers of the Industrial Revolution. Together these forces were changing life beyond recognition. The first zeppelin flight had just taken place over Lake Constance in

Germany and the first automobile show was opening in New York's Madison Square Garden. Seafaring was also being transformed. Britain's Royal Navy was preparing to drop its first submarine into the grey waters at Barrow-in-Furness. And for the first time, gleaming steamships traversing the world's oceans outnumbered vessels propelled by sail.

In the Mediterranean, the revolution had reached one of the most prominent local industries: sponge diving. Since well before Homer's time, Greek divers had earned a living cutting sponges from the seabed; we know that the ancients routinely used them for bathing and for cleaning the house. In one of the most famous examples, after the wandering hero Odysseus finally returns home to take violent revenge on the men who have been wooing his wife in his absence, he has his maidservants sponge the corpses' blood from the tables (before he hangs them, too, for their disloyalty).

The sponge divers' profession changed little over thousands of years, from perhaps 6000 BC, when the earliest signs of agriculture appeared on Greek soil and the first ships ventured out across the Aegean Sea. The most accomplished and daring divers came from the south-eastern Dodecanese islands, especially Kalymnos and Symi, where the warm water nurtures particularly large specimens. Naked and armed with a sharp knife, the athletic sponge fishers would dive to around 30 metres, weighed down by a large, flat stone, and collect sponges in a net for as long as their lungs would allow.

But, in the nineteenth century the sponge-diving industry was transformed for ever. Perhaps the change was inevitable, but if you're looking for a particular individual to pin it on you could say it was down to a rather distinguished German engineer called Augustus Siebe. After learning metalworking in Berlin and serving as an artillery officer at the Battle of Waterloo, Siebe settled down in Soho, London. A prolific inventor, he had among other things a rotating water pump, a paper-making machine, a weighing scale and an icemaker to his name. Then in 1837 he invented a diving helmet, fitted to a watertight canvas suit.

Like all Siebe's inventions it was quite ingenious, although this contraption was to have a far greater impact than any of his others. Thanks to a valve in the helmet, a diver wearing the suit could breathe air fed through a hose from a compressor in a boat above. For the first time divers could descend as deep as they liked, or as far as the air hose would reach anyway, and stay underwater for much longer periods. The potential economic benefits for sponge diving were huge and in the 1860s the new suits were brought to Symi by an enterprising local merchant called Fotios Masatoridis.

Each suit consisted of thick folds of canvas, tightly sealed with rubber and bolted onto a large bronze collar and breast-plate. Screwed on top was a round copper helmet, so heavy that it took two hands to lift it, and once imprisoned inside the diver had only little portholes made of reinforced glass to see through. It was impossible to swim in such armour. Instead the divers had to trudge along the bottom, dragging air hose and lifeline behind them, like primitive astronauts tethered to a hovering spaceship on some dense, high-gravity planet.

The expert divers were wary, to say the least, when confronted with these bizarre outfits. Then Masatoridis persuaded his pregnant wife to demonstrate. Obligingly clad, she clambered down the harbour steps until the waters met over her head. The helmet performed perfectly. Being upstaged by a woman – and a pregnant one at that – was unthinkable – so the suits were quickly accepted.

At first they seemed miraculous. After some practice, the divers routinely descended to 70 metres below the surface. There they could tramp around on the seabed, hunting for sponges and harvesting them at leisure, while communicating with the boat above by means of a string tied to one wrist. The vastly increased harvest transformed the industry and the merchants selling these bumper hauls (if not the divers themselves) made huge fortunes. At the industry's height, between 1890 and 1910, thousands of divers worked each year, putting in perhaps a million hours of time on the sea bottom between them.

There was a tragic human cost for this financial success, however, as the suits brought with them a prolific and indiscriminate killer: the bends.

If you breathe compressed air at depth, the nitrogen in the air in your lungs is at a higher pressure than it is in your body, so it dissolves in your blood and tissues until an equilibrium is reached. This isn't a problem, until you want to return to the surface. Then, if you go up too fast and the pressure drops too quickly, the nitrogen in your body doesn't have a chance to pass back into the air. Instead it pops out of solution as bubbles – just like the dissolved carbon dioxide does when you pop the cork on a bottle of champagne.

The symptoms of the bends depend on where the bubbles form – most commonly they appear in the joints, causing excruciating pain and leaving you unable to straighten your limbs. In the brain the bubbles cause confusion, memory loss and headaches. In the spinal cord and nervous system they cause paralysis; in the skin they cause itching, and a sensation of tiny insects crawling over the body. Bubbles can clog up your veins, cut off the spinal nerves or cause a heart embolism. Severe cases are fatal, and it's not a nice way to go.

The first cases of the bends were reported in the 1840s, not in divers but in miners and bridge construction workers who were exposed to underground shafts in which the air was kept at high pressure to keep out the water. The condition got its name from workers digging the pier excavations of the Brooklyn Bridge in the 1870s. They often came up in tortured body positions that reminded their colleagues – rather callously, it has to be said – of an affected curve of the back that was a popular pose among women at the time, known as 'the Grecian bend'.

But the sponge divers who started using the new diving suits in the 1860s didn't know any of this. It wasn't long before they started dying, and in huge numbers. Between 1886 and 1910 around 10,000 divers died from the bends and 20,000 were paralysed; that's about half of those who went out on the boats each year.

The impact on the sponge-diving communities was

enormous, with almost every family affected. Largely due to pressure from divers' wives and widows, the suit was soon banned in many countries, including Lebanon and Egypt. But a mixture of commercial pressures and pride kept the Dodecanese divers using it. Compared to a mundane life on dry land, diving gave them a chance for money and glory; as in war, every day was lived as if it were their last.

Now more than ever the sponge fishers were a tribe apart. Young, macho and proud, they faced great danger for the riches they brought home and were seen as glamorous heroes on the tiny islands from whence they came. Every spring, fleets of flimsy wooden boats would leave Symi and the surrounding islands, each carrying anything up to 15 divers who shared one battered suit and hand-powered air compressor. They would spend an exhausting summer living and working on the boat, travelling as far afield as North Africa. In the autumn those who survived would return, laden with cargo and ready to celebrate.

So it was that in the autumn of 1900, Captain Dimitrios Kontos and his crew were sailing home to Symi from their summer sponge-fishing grounds off the coast of Tunisia. Kontos was a former master diver himself, but was now in charge of two tiny sailing boats. Under his command were six divers as well as 20 oarsmen, so they could still make progress on windless days.

Their *caiques* or cutters were just a few metres long and built pretty much as sponge-divers' boats had been since before Homer's time – the outboard motor would not reach the

Aegean for another couple of decades. Vertical beams wedged tightly inside a horizontal wooden frame formed the delicate S-shape curve of the hull, while a spider web of ropes splayed down from the fragile masts, each proudly topped with a Greek flag. (Symi, along with the rest of the Dodecanese islands, remained under Turkish rule until 1923, but the inhabitants nevertheless saw themselves as fiercely Greek.) After six months' hard work the decks were so densely filled with drying sponges that there was hardly room to move, with yet more strung from every available inch of rigging.

The way home took Kontos and his men northeast from Tunisian waters and up to Cape Malea. But like so many sea travellers following the route before them they fell foul of a violent gale, and were blown towards a barely habited islet.

This island has been given many names in its long history. The ancient Greeks called it Aigilia, which the handful of locals later morphed into Sijiljo, while passing sailors who spoke the Italian-based lingua franca of the Mediterranean called it Cerigotto. These days, however, it is known as Antikythera (pronounced with the accent on 'kyth', to rhyme with pith). A lozenge shape just three kilometres wide, Antikythera sits 40 kilometres south of Kythera – and right in the middle of the passageway between Cape Malea and Crete. Centuries ago, Antikythera was covered in lush greenery, but the inhabitants cut down the forest to build ships. They couldn't have known the effect it would have. Without the tree roots to hold it in place the soil was gradually carried off by the incessant winds, leaving the island beautiful but barren.

Stormwaters around this deadly shard of rock are not for the faint-hearted. The sea turns almost black and angry waves attack the rocks; any ship unlucky enough to find itself in the way is likely to be deftly dashed to pieces. But Kontos was a skilled skipper and he managed to guide his men to shelter in the island's only harbour, a tiny cove on the northern coast called Potamos, where a handful of white houses are scattered like sugar cubes over the dark, rocky soil.

After three days the winds died, the waters returned to a smooth, glistening blue and the divers' thoughts turned to checking out what was beneath them. Always hoping to find late additions to their hard-earned haul, Kontos took one of the boats out around the sharp, rocky headland just to the east of the port, to a submarine shelf known by the locals as Pinakakia. He dropped anchor about 20 metres from the steep cliffs.

Elias Stadiatis was the first diver into the water that morning. He sank down quickly to the sloping shelf 60 metres below, but reappeared just five minutes later, clearly agitated. His comrades hurriedly hauled him aboard and twisted off his hefty copper helmet.

A huge mound of men, women and horses. Decaying, rotting. Must have come from a wrecked ship. Stadiatis breathlessly recounted what he had seen lying on the seabed. No part of the ship itself was visible – any wood exposed to the water would have long ago been devoured by shipworms. But its ghostly cargo was plain to see.

Kontos pulled the dripping suit off his gabbling friend

and donned it himself to investigate. After he had dropped through the cold water for a couple of minutes, a tumbled mass of figures, parallel to the shore and about 50 metres long, loomed out of the blue. They weren't corpses but statues – corroded and encrusted with marine sediment, yet for the most part clearly recognisable. Some were marble, while the shafts of sunlight penetrated just deep enough to reveal that others had a green tint: the tell-tale sign of ancient bronze. As his boots sank into the slanting mud and his air hose snaked up through the water to the dim shadow of the boat suspended far above, Kontos struggled to keep his breathing steady. This wreck had been carrying treasure.

He grabbed a bronze arm from one of the statues as proof of the find, attached it to his life line and headed triumphantly back to the surface.

Sources differ about what happened next. The official Greek version is that Kontos ordered his men to measure and record the location of the wreck, before they finally sailed home to Symi.[1] After enjoying the customary heroes' welcome, Kontos informed the island's elders of the find and asked them what to do. Full of patriotic pride, they recommended that he leave immediately to report the discovery to the Greek government in Athens.

1 The account presented in this text is based on the official report of the Archaeological Society of Athens, published shortly after the discovery in 1902, which states that the wreck was discovered "at the end of 1900", when the divers were on their way home. However the Symiote historian Eleni Kladaki-Vratsaniou suggested in 2007 that the divers may have discovered the wreck in spring, on the way to their summer diving grounds, but did not report the find until they returned in autumn.

But perhaps they weren't in such a rush. Peter Throckmorton, an American archaeologist, journalist and diver who was involved in excavations of several Mediterranean wreck sites in the 1950s and 60s, studied the Antikythera finds and interviewed people on Symi. Few who remembered the discovery were still alive, but stories of it were still eagerly told in the taverns along the seafront. According to Throckmorton, the locals' story was that Kontos and his men first used ropes to lift whatever they could from the wreck site for themselves before the weather changed that autumn. He points out that there are rumours of many small bronze statues being sold in Alexandria between 1902 and 1910, and that the lead bars from the ship's anchors have never been found. Lead would have been precious to the divers, for use as weights. When they could salvage nothing else with their tiny boats, the divers went to the Government in the hope of a reward.

Either way, at some point Kontos and Stadiatis, with the bronze arm in tow, went to see a Professor A. Ikonomu, an archaeologist at the University of Athens who came originally from Symi. On 6 November 1900 he took them to the office of the Minister of Education, Spyridon Staïs.

It was good timing. No archaeological survey of a wreck had ever been undertaken, in Greece or elsewhere, but the Government, led by Georgios Theotokis's New Party, had just begun to realise the potential of raising ancient riches from the seabed. Sixteen years earlier it had funded a survey to look for remains from the greatest sea-battle in Greek history, when the fleet of the Persian king Xerxes was crushed by the Greek navy in the straits of Salamis in 480 BC.

A few noteworthy items had previously been recovered from the sea, including a bronze chest protector found in Pylos harbour in southern Greece, ancient timbers and two life-size marble statues at Piraeus, Athens' port, and an inscribed lead anchor from the harbour at Symi. All had been discovered by chance, either by sponge divers or dragged up in fishermen's nets.

So it was quite a bold move when in 1884 the Archaeological Society of Athens, with the Government's backing, decided to go out and actively search for submerged artefacts. Modern Greece was a young, relatively insecure country, having escaped Turkish rule only in 1830, and the Government cannily thought that recovering the remains of past glories would do wonders for national pride. Unfortunately the society didn't know of any wreck sites to explore, so after much pondering it chose the straits of Salamis as its expedition site. In the frenzied battle there nearly 2,400 years before, the Greeks had lost 40 triremes (wooden warships, named after the three rows of oarsmen on each side), while the Persians lost a whopping 200. The sea floor was surely strewn with their remains.

It was a much harder mission than the archaeologists had bargained for. The water was only 20 metres deep, but bad weather meant the hired divers could only work twelve days out of the month that had been scheduled. Even on calmer days the choppy sea stirred up mud from the bottom so the divers couldn't see what they were doing. And in any case there was so much seaweed and clay everywhere it was impossible to tell what lay beneath. For a cost of precisely 1548.50 drachmas (worth around £8,000 today), the team came back with a few fragments of amphoras, a nearly intact vase and a wooden plank that broke as it was being brought up to the surface.

In a dejected report to the Archaeological Society later that year, Christos Tsoundas, who had supervised the expedition, described it as a 'complete failure'. Looking back, however, that was a little harsh. For the first time in history a professional archaeologist had led a team of divers, taken rough measurements, recorded what was found and reported back. The idea of underwater archaeological surveys carried out by the state had taken root, even if the disappointing outcome meant that no other projects were attempted for the next few years.

The expedition also caused a fair amount of excitement in the press at the time (before the final haul was known, at least) and it may have been the memory of this that encouraged Kontos and his elders to reveal their find at Antikythera to the Government. Officials at the education ministry initially met Kontos's claims with disbelief. No sunken ship had ever been found in Greek waters and the divers' story seemed too good to be true. But the evidence of the bronze arm, and the potential value of the find, won them over. This project promised to yield everything that the Salamis mission had failed to. According to Kontos's account, a wreck had already been located and the divers had already established that it contained treasures to be salvaged. If it had bronze statues on board, the ship was surely close to 2,000 years old, for no such artefacts were made after Greek civilisation fragmented in the early centuries AD, and any that survived (unless buried somewhere or out of reach at the bottom of the sea) were soon melted down as scrap metal.

If the Government would provide the necessary equipment to winch the sunken objects up from the seabed, Kontos told the minister Staïs, his men would dive for them – provided they were paid the full value of whatever they recovered. Slightly nervously, Staïs agreed to Kontos's terms, as long as an official archaeologist was on board to oversee the project. Professor Ikonomu was appointed and Kontos handed over the arm.

Staïs moved fast. Once the location of the wreck got out, looting was a strong possibility. And perhaps Kontos would change his mind. So within a few days a navy transport ship called the Mykale took Ikonomu to Antikythera, accompanied by Kontos, the divers and the oarsmen, in their two little fishing boats. After being slightly delayed by bad weather, they all arrived at the wreck site on 24 November. The divers – Ioannis Petiou, Giorgios and Kyriakos Mountiadis, Elias Stadiatis, Konstantinos Kalaphatis, Giorgios Kritikos, Basileios Zouroudis, Basileios Katsaris – began their work.

At the wreck site, the cliffs of Antikythera drop vertically to about 50 metres below the sea's surface. Then there's a shelf of sandy mud, on which the ancient ship came to rest, which slopes gently down to about 60 metres before dropping off again to deeper water. Ikonomu and Kontos had agreed a plan of action. Light objects from the sunken cargo were to be attached to ropes and raised using winches attached to the divers' own boats, and heavier ones were to be lifted with the sturdier hoist of the *Mykale*. But in that first run, the sea was still pretty rough. Swells from the north punched against the cliffs and it became clear that the *Mykale* was too large to get safely close to the rocks. Kontos, eager to prove the truth of the find and not one to be deterred by a little inclement weather, sent his men down anyway. In the three hours before the worsening storm forced them to stop they brought up a bronze head of a bearded man, the bronze arm of a boxer, a bronze sword, two small marble statues (both missing their heads), a beautifully crafted marble foot and several fragments of bronze and marble statues, as well as bronze bowls, clay dishes and other pottery.

Returning to Athens to be replaced by a smaller craft, the *Mykale* took these rewards home in triumph. Staïs must have breathed a huge sigh of relief when he realised that his investment had been a wise one after all. The divers really had stumbled upon the biggest hoard of ancient Greek bronzes ever found. The story became front-page news and, as the Government had hoped, all of Greece (but especially Athens) was set alight by a collective and patriotic excitement. After centuries of having their treasures looted by everyone from the Romans to the British, some antiquities were finally making it back home.

The navy assigned a more manoeuvrable ship, the steam schooner *Syros*, to the Antikythera mission and she arrived at the wreck site in time for the divers to start work again on 4 December 1900.

The conditions they faced were treacherous. Top of the list of difficulties was the unwieldy suit, which was not designed for the hard physical work of digging and lifting statues. To make matters worse the waters around Antikythera are cold and prone to sudden currents, as well as frequent gales and storms. The salvage expedition lasted ten months, until September 1901, yet the weather prevented the divers from working even a quarter of those days. For the rest, they had to sit out the storms on their tiny boats.

But the biggest challenge of all was the depth of the wreck, which was to push the divers to their limit. At about 60 metres down the site was well out of reach of any navy in the world at that time. Even by 1925, for example, just 20 US navy divers were qualified to dive to 30 metres. For Kontos's men to reach the wreck at all with the equipment they had, let alone do heavy work down there, was an incredible achievement. It's likely that no one but the Mediterranean's most daring sponge divers – who practically grew up in the water and depended for their livelihoods on going deeper than anyone else – could have managed it.

Although the divers at Antikythera had no comprehension of the diving tables or decompression stops used for safe diving today, or of what the depth was doing to their bodies, they at least realised that limiting the time they spent on the bottom reduced their chances of dying. They limited their submersions to five minutes on the bottom, twice a day, coming up reasonably slowly (meaning that between them, the six men could only work a total of an hour on the bottom each day). But even with the best intentions their diving suits were hard to control, especially when ascending. A diver had to carefully monitor the amount of air in his suit by balancing the air he vented with the intake from the valve in his helmet. If he miscalculated and allowed in too much air it would expand within the suit as he rose, carrying him helplessly ever faster towards the surface, and a sure case of the bends.

There was also a second depth-related danger to contend with – nitrogen narcosis. Well known to most scuba divers, it is a mysterious alteration of consciousness thought to be caused by the effect of high nitrogen pressure on nerve transmission. The French diver-explorer Jacques Cousteau famously described it as 'the rapture of the deep', because it feels as if you are giddily drunk. It gets worse the deeper you go, kicking in at around 30 metres and getting progressively more serious – budding scuba enthusiasts are taught to remember that it's like having one martini for every 10 metres below 20 metres. The effects reverse as soon as you ascend, but the impaired judgement they cause means that some divers never surface. The rapture makes them feel invulnerable and affected divers have been known to throw away their masks or swim far down to their deaths.

Cousteau's diving colleague Frédéric Dumas described the effects of nitrogen narcosis at 70 metres, just a little deeper than the divers at Antikythera had to reach, in Cousteau's 1953 book *The Silent World*:

> My body doesn't feel weak but I keep panting. The damned rope doesn't hang straight. It slants off into the yellow soup. I am anxious about that line, but I feel really wonderful. I am drunk and carefree. My ears buzz and my mouth tastes bitter. I have forgotten Jacques and the people in the boats. My eyes are tired. I lower on down, trying to think about the bottom but I can't. I am going to sleep, but I can't fall asleep in such dizziness.

That was in 1943, when scuba-diving gear had just been invented by Cousteau, together with Émile Gagnan, an expert in industrial gas equipment in Paris. Gagnan had been working on a demand valve to feed cooking gas into car engines. The Second World War had caused petrol shortages and everyone was trying to work out how to run cars on the fumes of burning charcoal and natural gas. Feeding compressed air into a diver's mouthpiece on demand turned out to be a similar problem.

By lowering themselves down a rope and signing their names on boards attached at various measured depths, Cousteau and Dumas wanted to see just how deep it was possible to go. Subsequent dives down to 100 metres set what is still sometimes called the 'theoretical limit' for diving with compressed air. It's more than a theoretical barrier, though. When the fearless experimenters tried to extend the limit with a 130-metre rope, their close friend Maurice Fargues was first to descend. At first all seemed to be going well, but after a few minutes his signals to the surface stopped. Cousteau and Dumas dragged him up by his lifeline; when he appeared beneath them they were horrified to see his body limp and his mouthpiece hanging from his chest. Twelve hours of desperate and exhausting resuscitation proved unsuccessful. When they later pulled up the marker rope, they found Fargues's illegible signature on the very lowest board.

For Kontos and his men in 1900 the exertion of hauling statues and artefacts, and attaching them to the winch line of the boat above, caused them to breathe heavily, making the effects of nitrogen narcosis even worse than normal for the depth they were at. Their helmets trapped the carbon dioxide they breathed out (unlike scuba gear, which releases the air you exhale into the water), befuddling them even more. Visibility was a problem, too, as the mud and sand floated up from the bottom in clouds as soon as the divers moved anything.

Yet throughout the winter the sponge fishers dived again and again and brought up find after find, while the archaeologists on board, dressed smartly as always in dapper suits, looked on. And despite the vile conditions the divers worked carefully – often taking several days to dig and clean around an object before easing it from the slippery mud. The supervising archaeologist George Byzandinos (who had taken over from Ikonomu) concluded that the fragments being brought up had broken thousands of years earlier, not in the salvage process. He even praised the divers for showing as great an interest in the preservation of the antiquities as any 'inspired fan of the old art'.

By Christmas their haul included plenty of marble statues, mostly of men or horses, another bronze sword, a bronze lyre, a colossal marble bull and various fragments of bronze furniture, including a throne. And there was the most exciting find so far: a beautifully worked bronze statue, perhaps of Hermes or Apollo. Although broken into several pieces it was quite well preserved and was hailed as one of the finest bronze statues to survive from antiquity. Perhaps it was even the work of one of the great fourth-century classical sculptors, Lysippus or Praxiteles, the archaeologists speculated excitedly. There was one statue that got away, though. The body of a large horse tore itself loose from its chains just as it reached the surface and crashed back into the sea, falling down the cliff into the deeper water beyond the divers' reach.

The statues piled up on deck and although the bronzes had survived pretty well, most of the marbles were terribly eroded. Emmanuel. Lykudis, one of the governtment representatives present, described the scene in his diary entry for 7–10 February 1901:

The sea has affected them in a terrible way. Most of them are now transformed into shapeless sea-rocks, and they have the appearance of tremendous sea shells . . . But under the transformation and destruction that the sea has carried out, one suspects the old glory, believes one can still recognise the beautiful lines.

The statues that had been recovered so far were transported back to Athens and put on public display at the National Archaeological Museum. Eroded or not, crowds flocked from far and wide to see such treasures from their nation's past, while the newspapers reported on every detail of the unfolding adventure.

Back at Antikythera, however, the storms, the punishing depth and gruelling regime of work were taking their toll, and the divers were beginning to suffer from exhaustion. By February, according to an account of the finds published in 1903 by John Svoronos, one of the country's most senior archaeologists, the men were often emerging from the water 'half-dead'.

The mood soured as the pace of discoveries slowed. Then the divers announced a further problem: part of the wreck, they said, was obscured by enormous boulders. After some discussion the archaeologists worked out that these must be rocks from the cliff above, dislodged at some time by an earthquake, and soon devised a strategy for shifting them.

They instructed the divers to dig tunnels underneath the boulders, then twine strong ropes around them several times, an arduous task that took more than 20 dives for each boulder. The other end of the rope was attached to the sturdy *Mykale* (brought out again from Athens for the task), which then steamed at full power towards the open sea. Once dislodged from the wreck, the boulders were to be released from the ropes, rolled down the slope and into the depths below.

It was a risky strategy, described by a jumpy Lykudis as 'one instant of large and probably justified fear!' If the rope snapped, the sudden shock might be enough to capsize the *Mykale*. Or worse, if the boulder remained entangled in the rope, its weight might drag the ship down with it. To avert the latter fate, several of the crew gathered around the rope where it was tied to the ship, ready with hatchets. Fortunately, their services weren't required and several boulders were successfully dispatched over the underwater cliff.

But then the minister Staïs, who was visiting, had a startling thought. What if the 'boulders' were actually colossal statues, so overgrown and corroded that the befuddled divers, working in the dim light of the wreck site, had failed to recognise them? He ordered the next boulder to be brought to the surface – at considerable further risk to the ship. After some tense moments there was cheering from the decks as it heaved into view through the clear water. It was a huge, muscular Hercules, complete with club and lionskin – eroded but still recognisable as similar in style to the world-famous Farnese Hercules, kept in the Naples Museum. Presumably, they preferred not to dwell on the statues that had already been rolled forever out of reach.

At this point the ill and exhausted divers demanded a month's break, at least until Easter. At first, Staïs encouraged them to continue for a few days more by promising to increase their reward, but eventually they were beyond persuasion and went on strike.

They donned their suits again in April, with ten divers instead of the previous six, but the easily recoverable objects had already been brought up, so in the first week the returns were meagre. Then tragedy struck. One of the divers, Giorgios Kritikos, surfaced too fast and died of the bends, leaving his family without a pension. Accounts from the time pass over this inconvenience with few details about what happened or even any detectable regret at his passing beyond its effect on the work at hand. Indeed, Staïs's response seems to have been to threaten to hire Italian divers who he thought would be more efficient.

He never acted on these words though, and the long-suffering Symiotes worked on into summer. As the months passed, the team was increasingly troubled by a seasonal wind that affects the area, called the *melteme*. Coming in from the north-east, it can whip up a harsh storm in minutes. The wreck site was completely exposed and it became harder and harder to work there. Every loose object had been removed from the wreck and the divers had begun to dig into the layers of sea growth underneath, but with little success. Then two more divers were seriously paralysed by the bends and work was finally suspended at the end of September 1901.

Those involved were reluctant to stop, believing that many more statues remained embedded there, especially as the bodies of many of the salvaged heads, feet and hands had not yet been found. From time to time the Government tried to hire foreign crews to continue the work, but these efforts failed because the divers wanted to keep some of the salvaged artefacts, which was forbidden by Greek law.

According to the official report of the Archaeological Society, the Greek government paid the 'conscientious citizens of Symi' – those who survived at least – a small fortune of 150,000 drachmas (equivalent to nearly half a million pounds today) as their reward, while the society paid them a bonus of another 500 drachmas each. The report proudly noted that the quality of the salvaged artefacts had exceeded all expectations.

It was a great success. The first ever archaeological survey of a wreck had yielded treasures far beyond what anyone had imagined. But the project was very different to anything that archaeologists might recognise today. There was no attempt to study the items from the wreck in context or to learn anything about the ship itself or the way of life on board. It was purely a salvage operation. None of the archaeologists would have dreamed of getting into the water themselves and they treated the divers as nothing more than hired labour. At no point, for example, did they ask them how the wreck and its contents were arranged.

Similarly, the finds were treated very differently to how they might be today. They were taken to the National Museum in Athens, under the supervision of the museum's director Valerios Staïs (Spyridon's nephew). But there was no effort to definitively catalogue the fragments and artefacts.

Some were placed on public display, but most were put straight into a rather jumbled storage. Inside the museum, a pretty open courtyard became the eerie resting place for heaps of marble statues, horribly disfigured by the action over the centuries of hungry sea creatures – everything from date mussels to marine bacteria. Men, women and horses were missing faces, heads or limbs, and their smooth, sculpted surfaces had been grossly eaten away, leaving sad, pitted shadows of the artist's original intent. Indeed, the crew members of the *Mykale* were so moved by one statue that was raised, of a beautiful but eroded young man, that they called it 'the ghost of Praxitelean Hermes'.

Every so often part of a statue was saved where it had been buried in the sand. One legless horse had a well-preserved body, and around a hole where the head would have fitted was a carved strip showing an eagle, helmet, Galatian shield and axe. Elsewhere a crouching boy, perfect on his right side with neatly cropped hair and eyes gazing upward, had suffered his left arm and leg being munched away to stumps.

The most valuable finds, though, were the bronzes. They were mostly in pieces, but the fragments, despite the metal on the surface having been corroded by electrochemical reactions with the seawater, had generally retained their original shape. Even so, most of the smaller pieces were thrown into crates and left in the courtyard, where museum workers would occasionally sift through them to look for bits that might fit the larger statues as they were reconstructed.

The big prize of the haul was a bronze of a naked young man, the beautiful Hermes or Apollo already mentioned. Nicknamed the 'Antikythera Youth', he stands calmly, nearly two metres tall, his right arm stretched out as if holding something. Although found in more than 20 pieces the statue was reconstructed in the early 1900s (then taken apart and reassembled again in the 1950s, with a slightly different posture) and its languid grace now presides over a central hall in the Athens museum.

Another colourful find was the portrait head of an elderly man, perhaps a philosopher, with piercing features, bushy beard and tousled hair. Plenty of other smaller bronzes were also found, in various poses, often decorated with eyes, nipples and genitals made of stone. One statuette of a naked young man was attached to a rotating base, presumably so that his slender form might be viewed from different angles.

Aside from the statues, many vessels and other small objects made out of clay, glass and metal were recovered. There were piles of amphoras (two-handled jars with pointed bottoms used for transporting supplies) of various sizes and shapes, one of them with olive pits still inside, as well as jugs, flagons, kettles, lamps, glass pots, bottles and a silver wine jar. A golden earring in the form of a baby holding a lyre was found alongside bronze bedsteads engraved with busts of a woman and a lion. And there were pieces of timber from the ship itself, as well as broken tiles from the galley roof.

The treasures from the Antikythera wreck still take up a large part of Athens' National Archaeological Museum. Since their discovery many other ancient wrecks have yielded precious finds, such as the first-century BC galley loaded with marble pillars discovered in 1907 off the coast of Mahdia, Tunisia, and a Greek ship from around 1 BC found at Cape Artemision with some of the finest ancient bronze statues known, from the classical height of the art in the fifth century BC. Other wrecks have been more scientifically excavated than Antikythera was, and can therefore tell us much more about the way of life on board, such as a Bronze Age ship which sank off Turkey's Cape Gelidonya carrying a cargo of copper ingots from Cyprus. But although many of the treasures from Antikythera have since been matched or even eclipsed, the ship from which they came retains its well-deserved place in history as the first wreck ever to be explored by an archaeological expedition, and the courageous efforts of the sponge divers remain as awe-inspiring today as they were in the 1900s.

That, though, is only the beginning of the Antikythera story. As the salvaged objects from the wreck arrived back in Athens, busy museum staff struggled to cope with the huge influx of artefacts, as they tried to piece together the larger statues and vases. So nobody noticed a formless, corroded lump of bronze and wood lying in one of the courtyard crates. But as the wood dried and shrivelled over the next few months, the secrets inside could contain themselves no longer. The lump cracked open to reveal traces of gearwheels embedded in the newly exposed surfaces, along with some faint inscriptions in ancient Greek.

2

An Impossible Find

To the antiquities brought up from the bottom of the sea at Antikythera belongs a completely strange instrument, whose purpose and use are unknown . . . It is nevertheless very similar to the gear wheel system of a simple modern clock.

— PERICLES REDIADIS

ONE HUNDRED MILLION years ago reptiles ruled a fiery Earth. Dinosaurs terrorised the land, while flippered icthyosaurs and plesiosaurs patrolled the oceans and pterosaurs fought with fast-evolving birds for domination of the skies.

All around them, the planet's crust was punctured by unusually violent volcanic activity. In fact, many believe that climate change caused by the flames eventually contributed to the dinosaurs' destruction, clearing the way for the rise of mammals. The impact of the giant eruptions was just as great beneath the sea. The mid-ocean ridge – an underwater mountain range formed where the Earth's tectonic plates join – was forced open by the strong currents in the hot mantle beneath. Seawater crept down into the cracks and mixed with the molten rock, dissolving minerals from it before being ejected from hot springs that burst up through the ocean floor; the pressure at such depth keeping the water liquid despite temperatures of hundreds of degrees.

The huge scale of this activity enriched the oceans with calcium, which plankton use to build their intricate skeletons of calcium carbonate. As millions of generations of plankton lived and died, more calcium carbonate was deposited in the oceans than ever before, forming the extensive beds of chalk (the White Cliffs of Dover included) from which this period, the Cretaceous, got its name (*creta* is Latin for 'chalk').

The superhot springs also carried dissolved sulphides of iron, copper, zinc and nickel. These precipitated out as dark solids as soon as they hit the cold seawater, forming angry black plumes of undersea smoke, which formed deposits of ore on the seabed wherever they settled. Normally these deposits were recycled back into the mantle beneath – the rock of the deep ocean bed is denser than the continental crust that forms the land and shallow seas of our planet, so when tectonic plates clash and the oceanic crust meets the continental crust, it's the oceanic rock that gets pushed down. On occasion, however, a piece of this ancient seabed peeled away and landed on top of a piece of continental crust, where over time it was forced up into the mountains.

And so a piece of oceanic crust from the ancient sea that separated Europe and Africa in the Cretaceous now forms the picturesque Troodos Mountains of Cyprus. Jump forward to just 5,000 years ago: the dinosaurs are long gone and people have inherited the gifts of the Earth. The islanders of Cyprus have learned how to smelt copper from the rich stores of blue-black sulphide ore they find in the forested slopes, using it to make tools, jewellery and weapons. Later they will make a fortune selling copper to traders – Phoenician, Greek, then Roman, who will carry ingots of this valuable metal far across the Mediterranean Sea to Greece, Italy, Asia Minor and Egypt.

The mountains of Cyprus are young upstarts, however, compared to the wise old rocks of Cornwall, south-west England. The granite of Cornwall cooled during the Devonian period, around 400 million years ago, when the first fish evolved legs with which to crawl on to the land, and the first ammonites and trilobites colonised the sea. Before the granite completely solidified, hot magma bubbled from below and forced itself up into vertical fissures in the cooler rock. The minerals in the magma crystallised as they cooled, leaving behind lodes of colourfully named and valuable ores, including wolframite, chalcopyrite, sphalerite, galena and especially cassiterite – an oxide of tin.

Jump forward again and, as with Cyprus, traders from all over the Mediterranean come to Cornwall to purchase the smelted tin. They collect it at the tiny rocky island that will one day be called St Michael's Mount and carry it back to France, where they load it on to horses and travel overland for 30 days until they reach the mouth of the Rhone, where ships will continue the journey to the sea.

Although copper was at first widely used for weapons and utensils, it was quite soft and easily dented. A copper axe would not keep its edge for very long and a copper shield would soon wear down against sharpened stones. Then, probably in several different places at several different times, someone realised that adding a small proportion of tin to the copper, around 10 per cent, made it stronger and harder. It also lowered the melting point of the new metal, giving smiths more time to cast it as it cooled. Knowledge of this new alloy – bronze – spread around the world, ushering in the Bronze Age, which in the Mediterranean began around 2500 BC.

Advanced metalworking techniques were developed and complex networks of trade, supported largely by the markets in copper and tin, stretched from Africa and Asia Minor to the north of Europe. The glint of bronze was everywhere and the people of the Mediterranean were richer than ever before.

It didn't last. Somewhere around 1200 BC everything collapsed. Why this happened is one of the most controversial questions in ancient history – theories put forward include economic decline, climate change, earthquakes and invasion – it may even have been a combination of all of these. Whatever the cause, trade stopped, kingdoms fell apart, skills such as navigation, metalworking and literacy were lost and the region was plunged into a dark age from which very few records survive. Without the trade networks to unite copper and tin, bronze was hard to get hold of, so iron – though not as strong or beautiful as its rosier cousin – became the metal of choice for weapons and other implements.

By Homer's time, around the eighth century BC, Greek civilisation was pulling itself out of the darkness, with the rediscovery of old skills and the rise of city states such as Athens and Sparta. By this time craftsmen had worked out how to add carbon to iron to make steel, which is much harder than wrought iron. But bronze was still prized – objects made of it were never discarded, but melted down and used again and again over generations. A dagger that lost its point might become beads or a bangle, then be sold and recycled into a cooking pot or a bedstead suitable for a royal household, then reincarnated as a chariot wheel or a statue, a knife, an axe or a spearhead.

At some point, however, a piece of worn out bronze was melted down and recycled not into any of these items, but into the delicate gear work of a complicated scientific mechanism. And because of a quirk of fate – a ship in the wrong place at the wrong time – this particular mechanism was never melted down. Instead, it sank 60 metres to the sea bed off Antikythera and lay there until Captain Kontos and his divers retrieved it at the turn of the twentieth century.

Bronze fares a lot better than many other metals when languishing for long periods under the sea. Seawater is a soup of charged ions – mainly hydrogen and oxygen from the water and sodium and chloride from the salt, but there are others floating around too, such as sulphate and carbonate.

These ions do their best to attack any material they come into contact with. Iron, for example, oxidises completely in contact with sea water, losing its original form and eventually taking on the consistency of chocolate.

Copper, on the other hand, is relatively unreactive. The ions in the seawater strip electrons from any exposed copper atoms, forming positively-charged copper ions that each react with a negatively charged chloride ion to form copper chloride. Similarly, tin reacts with oxygen ions to form tin oxide. Certain marine bacteria do their bit too; their idea of a satisfying meal is to combine sulphate ions in the sea water with metal ions to form tin and copper sulphide, releasing chemical energy in the process. But the damage is limited. These new compounds form a thin layer on the surface of any bronze object that is left in the sea, which protects it from further corrosion.

That's why the bronze statues brought up from the Antikythera wreck were quite well preserved – once they were cleaned, the original form of the ancient figures was revealed. But the chemicals formed by the corrosion of bronze can turn nasty. Copper chloride is a stable compound in water, but not in air. When objects that have been corroded in this way are removed from the sea, the copper chloride reacts with oxygen and moisture in the air to form hydrochloric acid. The acid attacks the uncorroded metal beneath to form more copper chloride, which again reacts with air to form more acid, and the cycle continues. If the reaction isn't stopped, the object slowly and inexorably self-destructs.

For many months before it was discovered the mysterious Antikythera mechanism sat in a crate in the open courtyard of the National Archaeological Museum in Athens, unnoticed, untreated and literally eating itself away. By the time an unnamed museum worker noticed the significance of the decaying, fractured lump and brought it to the attention of the museum director Valerios Staïs, the outer layers of bronze had been completely destroyed.

Shrivelled fragments of wood clung to the bronze pieces, suggesting that the object had once been housed in a wooden box about the size and shape of a squat dictionary. Perhaps as the water evaporated out of the wood, the force of shrinking had literally pulled the contents apart. Perhaps a museum employee, eager to see what was inside, had hit the lump with a hammer. Either way, it was now in four crumbling bits.

An ugly layer of limestone – mostly calcium carbonate deposited as sea creatures feeding on the wreck died – covered much of the outer surfaces. But where the lump had cracked open patches of colour revealed the army of reactions devouring the bronze. The whitish green and bright blue-green of different forms of copper chloride dominated, but snaking through the green Staïs saw streaks of brownish-red copper oxide, the brown-black and whitish grey of various forms of tin oxide, and even the yellow and blue-black of tin and copper sulphide. Although a small core of metal remained in the centre, the surface of the fragments was made up of a powdery material that fell away at his touch.

Staïs was enjoying an impressive career. He was originally from the rugged island of Kythera, just north of Antikythera. Like his Uncle Spyridon, the Education Minister who first received news of the Antikythera shipwreck from Captain Kontos, Valerios had travelled to mainland Greece as an ambitious young man. He studied medicine, then archaeology, and he became director of the prestigious National Archaeological Museum of Athens aged only 30, just in time for the completion of the museum's first permanent home in 1889. Since then the new buildings had been filled with ancient Greek statues, tools, weapons, pots, jewellery, and not least the fabulous finds from Antikythera, which, over the past few months – incredible, chaotic, wonderful months – had gained international fame for him and his museum. But with all the precious artefacts that had come through the museum's doors, Staïs had never seen anything like this.

It was clockwork. Ancient clockwork. The largest piece of the strange object was about as wide as a page in a book, and almost as tall. One corner might once have been square, but the other sides were rough and eroded. Limestone formed an uneven layer over much of the front, although through it the outline of long-buried yet modern-looking gearwheels could still be made out. The overall effect was eerie and otherworldly, like finding a steam engine on the ancient, pitted surface of the Moon.

The clearest structure was a large wheel, almost as wide as the fragment itself, with a square hole in the centre where an axle might once have sat. Triangles had been cut out of the middle of the wheel, so that four unequally broad spokes formed the shape of a cross. Around the edge were about 200 tiny jagged teeth, carefully cut into triangles by some ancient hand and so small that they could only be counted with the help of a magnifying glass. A second, smaller, toothed wheel on the same side looked as if it might have engaged with the first, and there were hints, harder to make out, of other much smaller wheels or circles.

On the other side of this largest fragment, several more cogwheels were visible, with yet more little teeth – freshly revealed where the object had broken open and stunningly sharp and precise. Two medium-sized wheels lay one on top of the other, the upper one slightly offset from the lower, and several much smaller wheels were also visible, as well as a square peg. A thin, flat sheet of bronze appeared to be stuck over the bottom right-hand corner and it carried the remains of a Greek inscription. Strings of capital letters in a miniature, precise hand were so worn as to be practically unreadable, but they tantalisingly filled line after line without a single gap, as if the message had been too urgent to afford any pause between words.

A second slightly smaller fragment also had a flat sheet stuck on to one face, engraved with another inscription. On the back of it had been cut a series of concentric circles, which looked as though they might have served as guides for a rotating pointer. Rocky deposits completely covered the front of the third fragment, but on the back of it was part of an illegible inscription, as well as a raised ring that intersected with another raised, curved edge. An inscribed letter 'T' was just discernible inside the ring, and something that looked like a movable pointer projected from the centre. The surface of the fourth fragment was completely eroded, but from the size and shape it looked as though it might contain a lonely cog.

The number of gearwheels and the precision with which they were cut, along with the presence of various scales, pointers and inscriptions, perhaps instructions, immediately suggested to Staïs that this was a mechanical device for making accurate measurements or calculations.

But it couldn't be. The pieces crumbling in his hands had to be 2,000 years old and nothing like that had ever been found from antiquity. The ancient Greeks (or anyone else around at the time) weren't supposed to have had complex scientific instruments, or even, according to many scholars, any proper science at all. And clockwork wasn't supposed to have been invented until the appearance of, well, clocks, in Medieval Europe more than a thousand years later.

It's hard to overestimate the uniqueness of the find. Before the Antikythera mechanism, not one single gearwheel had ever been found from antiquity, nor indeed any example of an accurate pointer or scale. Apart from the Antikythera mechanism, they still haven't.

Ancient texts reveal slightly more, although with written descriptions it's harder to tell how the objects being described actually worked, or whether they were ever made. Also you often have to rely on texts in which the writer is describing something long after the event, or texts that have been copied and recopied many times and therefore could have been corrupted. But there are a few scattered mentions of gearing. The earliest may be a treatise on mechanics dating from around 330 BC, dubiously attributed to the revered philosopher Aristotle. It discusses circles that roll in contact, pushing each other round in opposite directions. The author might be talking about gears, but there's no mention of cogs or teeth, so it's hard to know for sure.

The first Greeks we know of to use working gears were the two most famous inventors of the third century BC: Ctesibius and Archimedes. Ctesibius, the son of a barber, became the greatest engineer of the time that we know about, after the legendary Archimedes, and he worked in Alexandria – in fact, he was probably the first director of the famous museum there. None of Ctesibius' writings survive, but we hear a lot about him from later authors, such as the Roman architect Vitruvius, writing a couple of hundred years later. Vitruvius said that Ctesibius built a water clock in which a float that rose with the water level moved an hour pointer by means of a 'rack-and-pinion' gear. This is a set-up in which a single gearwheel engages with a flat, toothed rack, and it's used to convert linear motion into rotational motion or vice versa.

Archimedes lived in the rich city state of Syracuse on the island of Sicily, although during his youth he almost certainly worked in Alexandria with Ctesibius. Among many other things he is credited with the invention of the wonderfully named 'endless screw' – in which a threaded screw is used to engage a toothed wheel with a much larger gear ratio. One full turn of the screw only turns the wheel through one tooth's worth of rotation – meaning that a lot of gentle winding turns the wheel only a small distance, but with a much stronger force than that originally applied by the winder. According to the ancient historian Plutarch, such a device allowed Archimedes to impress Syracuse's king by single-handedly dragging a ship over the ground, 'as smoothly and evenly as if she had been in the sea'.

Another, slightly more complicated device described by Vitruvius was a distance-meter or odometer, based on the principle that chariot wheels with a diameter of about four feet would turn 400 times in one Roman mile. For each revolution a pin on the wheel's axle engaged a 400-tooth cogwheel, moving it around the equivalent of one tooth, so that the cogwheel made one complete turn for every mile. This wheel engaged another gear with holes along its circumference that held pebbles, so that as the gear turned the pebbles dropped one by one into a box. Counting the pebbles therefore gave the distance travelled, in miles.

Perhaps Roman chariot drivers charged by the mile. We don't know for sure that the devices were ever built, but the general idea seems sound enough and they may have been around much earlier than Vitruvius' time. Alexander the Great was accompanied on his campaign in Asia in the fourth century BC by 'bematists', who had what must have been one of the most boring jobs in the ancient world – counting their steps to measure distances. Their accuracy even over journeys of hundreds of miles (they were often less than 1 per cent out) has led to suggestions that they must have had mechanical odometers to help them.

The height of invention in ancient Greek gears was supposedly reached by the instrument-maker Hero, another follower of Ctesibius and a lecturer at the Alexandria museum some time later in the first century AD. Hero wrote about the principle that Archimedes had started to develop of using gearwheels of different sizes to change the strength of an applied force.

In particular, he talked about a weight-lifting machine called a *baroulkos*. He drew a picture of it, showing how a series of gearwheels of increasing size would allow a relatively small force to lift a heavy weight. There's no evidence that it was anything but an armchair invention – indeed many scholars have argued that the teeth wouldn't have been strong enough for the device to work in practice – but the description alone proves that the principle of intermeshing gearwheels was understood. Another device described in detail by Hero was an elaborate *dioptra* or sighting instrument, which used an endless screw and cogged semicircle to allow it to be aligned accurately.

So we know that the Greeks used toothed gearwheels in simple mechanical devices from around 300 BC onwards. But most of these devices involved just one or two wheels that engaged with a screw or rack and they didn't need to be particularly precise, they just needed to apply force or lift a weight. Even so, Hero has been seen as an aberration in the history of technology: a genius who did not represent his age but described mad, impossible devices far beyond the comprehension of his peers. One eminent publication on the subject from the 1950s describes Hero's *dioptra* as 'unique, without past and without future: a fine but premature invention whose complexity exceeded the technical resources of its time'.

But compared to the *baroulkos* and *dioptra* – which were supposedly so far ahead of their time – the gearing in the Antikythera mechanism was undoubtedly real, and its complexity was breathtaking. These were precisely cut bronze gearwheels, clearly meant for some mathematical purpose. It was hard to count the gears embedded in the battered fragments, but to Staïs and his colleagues at least 15 wheels were visible on the eroded surfaces alone. They seemed to have interacted to make certain numerical calculations, the answers to which would have been displayed via pointers on a scale.

Rather than anything the ancient Greeks were supposed to have built, the sophistication of this mechanism made it look more like a clock or calculator. But if so it had to be nearly 2,000 years ahead of its time. Mechanical clocks of such a small size required delicate springs and regulators and didn't appear in Europe until the fifteenth century, and the first mechanical calculators – complex contraptions that used metal gearwheels to add, subtract, multiply and divide – weren't devised until some 200 years after that.

Today we're so used to electronic computers and calculators that the idea of a calculation using metal gearwheels might seem bizarre. Imagine, for example, that you have a gearwheel with 20 teeth that engages with a gearwheel with 10 teeth. Each time you turn the first gearwheel through one complete revolution, the second one will turn twice. In other words, your input has been multiplied by two (actually the second wheel turns in the opposite direction to the first, so you could argue that the input has been multiplied by minus two, but you get the picture). This is part of what clocks do – converting the seconds that tick by into the minutes and then the hours of the passing day. The more gearwheels you have, in series or in parallel, the more complex the calculation you can make.

The idea that an ancient clock or calculator might have been found caused excitement and some consternation at the Athens museum. Realising that interpreting it was beyond his expertise, Staïs quickly called in two experts. The first was John Svoronos, director of the National Numismatic Museum in Athens – the keeper of the nation's ancient coins and an expert in reading ancient lettering. Svoronos was one of the most senior archaeologists in the country and hugely knowledgable; unfortunately, he was also prone to coming up with eccentric theories about which few dared to disagree. The second expert was Adolf Wilhelm, a young and brilliant Austrian expert in inscriptions, who was stationed in Athens at the time.

Over the next few days Wilhelm cautiously dated the

writing on the mechanism to somewhere between the second century BC and the second century AD. Meanwhile, Svoronos and some of Greece's top scholars exchanged rival and rather pompous articles in the national press, hotly debating what the bizarre instrument might be – their talk of cogs and scales appearing alongside reports of Cuba's newly won independence from the United States and Britain's violent takeover of South Africa. Then the initial excitement died down, as the experts each went away to write up their various theories for scholarly publication.

Svoronos got there first, in a 1903 report written with Pericles Rediadis, a professor of geodesy and hydrography (fields concerned with measuring the Earth and the sea). Rediadis, a senior member of the Archaeological Society of Athens, was also interested in naval history and he was well known for his studies of the site on which the famous Battle of Salamis (480 BC) was fought.

Svoronos pored over the cryptic inscriptions on the Antikythera mechanism with a magnifying glass. He was able to decipher 220 scattered Greek letters, though very few whole words, and he compared their style to those on the ancient coins that he knew so well. He overruled Wilhelm's opinion on the mechanism's age and announced instead that the writing dated from the first half of the third century AD, a turbulent time of civil war when the Roman empire, including Greece, was ruled by a succession of leaders who each briefly seized power before being brutally assassinated.

Meanwhile Rediadis provided a description of the Antikythera fragments – the first technical, if rather vague, account of what he called 'this completely strange instrument'. He noted that the mechanism had been carried in a wooden box, as nautical instruments on ships still were in his own time, and deduced that the object was not part of the Antikythera ship's cargo, but a navigational instrument used by the crew.

From the scraps of lettering deciphered by Svoronos and Wilhelm, Rediadis suggested that the inscriptions were operating instructions, and put great importance on one particular and very unusual Greek word: μοιρογνωμονιον. This is a technical term referring to a graduated scale. The word was used to describe the zodiac scale in the earliest known account of the astrolabe, written in the sixth century AD. Svoronos and Rediadis concluded that the Antikythera mechanism must therefore be a kind of astrolabe.

Astrolabes were among the cleverest instruments thought to have been around in antiquity, and they were calculators of a sort. They were used for solving problems relating to the time and the position of the Sun and stars in the sky, and they were popular until the seventeenth century or so, when increasingly accurate clocks and astronomical tables began to render them obsolete.

The essence of an astrolabe, however, was something that the new technologies could never replace. The name means 'star catcher' and it is apt: holding the engraved, metal circle of an astrolabe you have the whole of the heavens in the palm of your hand. From Aristotle's time onwards, it was accepted (with just a few dissenters) that the Earth lay motionless at the centre of the universe, with the Sun circling around it and a sphere of fixed stars rotating behind that. An astrolabe is a flat disc in which one circular plate rotates over another to represent in two dimensions the spinning heavens as seen from Earth. The Sun, stars, horizon and even the very sky itself are represented by intricate patterns on its face. The inscriptions look complex and alien to us today, but they are the result of centuries of astronomical observations, and they elegantly encode our place in the visible universe.

The circular instrument's base – called a *mater* (or 'mother') in Latin – had a central pin over which a flat metal plate fitted snugly, like a disc on a record player. Engraved on it was a bewildering yet beautiful set of intersecting curves, lines and circles. This was a map of the sky, imagined as a sphere and projected on to the flat disc with the North Pole in the middle – just as a map of the Earth represents the spherical surface of the planet on a flat piece of paper. Although the sky looks like a featureless expanse, you can actually draw lines that mark very specific locations on it. The plate was engraved with a straight vertical diameter to show north and south, for example, and a horizontal diameter to show east and west. A series of curves and circles depicted the celestial equator (the Earth's equator as extended straight up into the sky), the Tropics of Cancer and Capricorn, and the horizon, as well as marking various altitudes above the horizon and degrees from north. The position of these lines is dependent on how far north or south of the equator you are, so most astrolabes came with a series of interchangeable plates, each engraved for a specific latitude.

Placed on top of this fixed sky map was a rotatable plate called a *rete*. This disc had key star constellations marked on it, as well as a circle to represent the path that the Sun follows through the sky. The whole sky rotates as the Earth turns, of course, but the Sun (because we're going around it) appears to us to move slightly slower than the stars through the sky, trailing them by a few degrees each day. The path the Sun traces with respect to the stars over the course of a year is called the ecliptic, because the only time you can see directly where the Sun is in relation to the background stars is during an eclipse. In ancient times, the 360-degree circle of the ecliptic was divided into twelve 30-degree sections of longitude, which correspond to the twelve signs of the zodiac. These were marked around the circumference of the astrolabe – this was the scale referred to in the sixth-century text, and in the inscription on the Antikythera mechanism.

On the *rete*, the spaces between the constellations were cut out so you could still see the sky map beneath (hence its name, which means 'net' or 'web'). The precise positions of the stars were represented by pointers, often in the dramatic form of flames or daggers, so that as the skeletal *rete* was rotated over the sky map, it showed the movement of the stars through the sky. Then on top of the *rete* was attached a rotatable straight bar, called a rule, which represented the Sun. The Sun's precise position on the sky map was given by the point at which the rule crossed the circle of the ecliptic. First the rule was set with respect to the ecliptic to show a particular day of the year, then it was rotated along with the *rete* to simulate the Sun's movement through the heavens on that day. Extra hour lines engraved on the fixed *mater* beneath allowed an astronomer to use the rule to read off the time at which the Sun or any marked star would hit a particular altitude.

Astrolabes were generally used for astronomical predictions and observations (there were sights on the back for measuring the altitude of stars or the Sun). They weren't especially helpful for navigation. Quite apart from the fact that the heavy metal disc would swing about clumsily in the wind if you tried to use it on deck, there were other, simpler devices for measuring the Sun's noon altitude, which was all you needed to determine the latitude of a ship at sea. And astrolabes did not measure longitude – how far east or west you were. There wasn't a way to do that until the eighteenth century, when the legendary British clockmaker John Harrison perfected clock mechanisms robust and accurate enough to be taken to sea to keep a record of the time at the ship's home port, and therefore show the time difference between that and the hour at the ship's present location, as indicated by the stars.

Although the first known description of an astrolabe comes from the sixth century and no actual instruments survive from before the ninth century, they were almost certainly around much earlier. The Greek astronomer Ptolemy described the maths necessary to make an astrolabe in the second century AD and reported lots of astronomical observations that were probably made using one. There's a colourful (if unlikely) story that Ptolemy invented the instrument when he was riding on a donkey, wisely pondering his celestial globe. He dropped the globe and the donkey stepped on it, squashing it flat, and giving him the idea. There are hints in other manuscripts, however, that the astrolabe may have been invented by Hipparchus, an astronomer who lived and worked on Rhodes in the second century BC, and from whom Ptolemy took much of the astronomy that he wrote about.

Svoronos and Rediadis's discovery of the zodiac scale certainly suggested that the Antikythera mechanism had something to do with astronomy. But it wasn't like any other astrolabe that was known at the time. For a start, astrolabes weren't square and they didn't come in wooden boxes. More fundamentally, although astrolabes had scales and pointers, they didn't have any need for gearwheels.

Like everyone else who saw the mechanism, Professor Rediadis was astonished at the sophistication of its gearing, and despite Svoronos's relatively late dating of it to the third century AD, he struggled to believe that this wasn't a much more recent instrument. To Rediadis, the gearing of the Antikythera mechanism looked just like the workings of a modern clock. If it wasn't for Svoronos's assurances that the instrument dated from centuries before the invention of the springs, regulators and escapements necessary for the continuous motion of a clockwork clock, he said, he 'would be bent to seize the history of nautical stopwatches from [John] Harrison'.

But when it came to identifying the mechanism, Rediadis was undaunted by its lack of similarity to known astrolabes. He thought that, as in a conventional astrolabe, the ancient instrument would have used a sighting line in combination with a degree scale to measure the height of stars or the Sun in the sky. But rather than using engraved maps and scales to read off the time of the day, say, or the longitude of the Sun, he speculated that the Antikythera device was a completely new type of astrolabe that calculated these values mechanically using trains of gearwheels and showed the result by means of the pointers. Although he called it an astrolabe (a description that would 'stick like a barnacle', as one historian put it, for the next half-century), he was really describing a sort of clock-like mechanism that instead of running automatically after being wound, was rotated by hand and set according to the movements of the stars.

From the sparse clues offered by the Antikythera fragments it was an inspired and quite beautiful guess. Unfortunately, neither Rediadis nor Svoronos addressed the question of why anyone would bother to build such a complicated mechanism to do what an ordinary astrolabe could have done perfectly well.

In 1905 another naval historian called Konstantin Rados,

who, like Rediadis, was an expert on the Battle of Salamis, published a paper arguing that the Antikythera mechanism was far too complex to have been an astrolabe. He, too, likened the gearwork to that of a clock and even thought he could see the remains of a metal spring in one of the fragments. Might this, after all, have been a mechanical clock capable of being wound? Rados could not believe that such a sophisticated device could have existed on the same ship as the ancient Greek statues recovered from Antikythera. He suggested that it must have dated from a second, much later shipwreck, and had found itself among the older remains by chance.

Two years later a young German called Albert Rehm entered the fray. He would go on to become one of the world's greatest experts on ancient inscriptions. But at this point he had just accepted a post at the University of Munich and was still making his name. Scornful of the lack of technical detail in Rediadis' description, and of the poor quality of his photographs, he went to Athens to examine the fragments himself, after which he sided with Rados and concluded that although it was certainly ancient, the mechanism could not possibly have been any kind of astrolabe.

By this time the fragments were being carefully, though controversially, cleaned. The treatment was revealing new markings and was necessary to prevent further corrosion of the bronze, but at the same time it destroyed some of the outer details. As a result of the cleaning, however, Rehm was able to read on the front dial of the third fragment a previously hidden and crucial word: *Pachon* (ΠΑΧΩΝ). Pachon is the Greek form of a month name in the ancient Egyptian calendar. There would be no use for the names of months on an astrolabe, Rehm argued, nor on any kind of navigational instrument.

He suggested that the fragments might be the remains of a planetarium. As a handle was turned, the differently sized gearwheels might have converted the motion into the appropriate speeds for each of the planets known at the time – Mercury, Venus, Mars, Jupiter and Saturn – showing their approximate paths as seen from Earth throughout the days, weeks and months of the year.

A ruffled Rediadis got his own back in 1910. In a new paper he argued that even if the mechanism wasn't an astrolabe, then a planetarium was a much less reasonable assumption – the gearwork was much too weak and flat for such a spherical device. He repeated his somewhat dubious argument that because the object was found on a ship and it had been housed in a wooden case, it must have been one of the ship's instruments.

After this, work on the mechanism stalled, despite the continued bickering of some of the world's most eminent science historians. The only major new research on the fragments around this time was done by John Theophanidis, a rear admiral in the Greek navy, who became interested in the mechanism in the 1920s when he was researching an article for a nautical encyclopaedia about the voyages of St Paul, who sailed back and forth across the Mediterranean preaching Christianity in the first century AD, before being shipwrecked on Malta while the Romans were taking him as a prisoner to Rome.

Theophanidis published his findings in 1934. As the limestone was scraped away, a large ring had been revealed on the front face of one fragment of the mechanism, with a graded scale around its circumference. Could this be the zodiac scale referred to in the accompanying inscription? Theophanidis also confirmed that the big cross-shaped gearwheel drove the rotation of several smaller gearwheels and he described a crank at the side that seemed to have driven the main wheel – wound by hand, Theophanidis suggested, or perhaps even driven by a water clock.

He also noted that the letters were so precise they must have been engraved not by a labourer but by a highly trained craftsman. Like all of the experts studying the device who came from a naval background Theophanidis became convinced that the mechanism was a navigational instrument. The inscriptions were instructions or rules, he concluded, which the ship's captain would have had copied for his personal use.

Theophanidis thought, like Rehm, that the device was for calculating the precise positions of the Sun, Moon and planets, with the ratios between the gear teeth producing their appropriate speeds of movement. But he couldn't quite give up on the astrolabe idea. In some of the engraved numbers he saw ratios reminiscent of the lines and circles of an astrolabe, and suggested that the inscriptions were instructions for tracing these markings with a ruler and compass, so that they could be used in conjunction with the instrument to solve various astronomical and nautical problems. He also speculated that by setting various pointers on the device according to the shadow cast by a nail placed in the middle of the concentric circles, it could calculate, by means of the gearing, the precise orientation of the ship.

Theophanidis became quite obsessed by the Antikythera mechanism and ended up spending many years working on his photographs of the fragments and building a model of the gearwork – to the extent that he had to sell several buildings that his family owned in the centre of Athens in order to finance his studies. But he did not publish on it again. Most of his extensive work lay unrecognised, hidden after his death within dusty piles of papers at his family home.

In the meantime, Albert Rehm's career went from strength to strength and in 1930 he was appointed rector of the University of Munich, making him one of the most influential academics in the country. But around him, everything was changing. The Nazi party had been growing in power since the mid-1920s, aided by a severe economic depression. Rehm was horrified to see the Nazi movement gaining ground among his students and did everything he could to dissuade them, without much success. After Hitler gained power in 1933, many of Rehm's Jewish colleagues had no choice but to flee the country. Rehm himself continued to protest vocally, earning the increasing displeasure of the regime, until he was forced to retire in 1936.

Nine years later, when the Second World War was over, Rehm was appointed rector once more in recognition of his resistance to the Nazis. But it didn't last long. He was just as outspoken against the new authorities for not recognising the importance of classical studies in German education, and he was removed from the position again in 1946.

Such stubbornness ran through everything that Rehm worked on and, like Theophanidis, he was unable ever to give up thinking about the ancient gears of the Antikythera mechanism. After his first paper on it he studied the fragments on and off for the rest of his life, intending to solve the workings of the device beyond doubt so that he could silence his critics with one triumphant, definitive publication. But despite his achievements in other fields, the secrets of the mechanism eluded him and the final paper never came. He died from a heart attack after attending a faculty meeting against his doctor's orders in 1949.

While Rehm had been fighting the Nazi regime, Hitler's shadow had reached Athens too. In April 1941 German forces advanced on the city and as the king and the government fled for Crete (except for Prime Minister Alexandros Koryzis, who shot himself in despair), the National Archaeological Museum was closed down. The precious exhibits were taken from their cases and buried in boxes – some in caves in the hills around Athens, some in the underground vaults of the Bank of Greece, and the rest under the floors of the museum itself, where they were hurriedly covered with sand. There the artefacts waited out the long, dark years of occupation, hidden from the looting army. Unfortunately there was no similar way to safeguard the city's food. The German soldiers, who had neglected to bring their own supplies, seized all they could from Athens's warehouses. By the time the buried exhibits saw daylight again, tens of thousands of Athenians had starved to death above them.

Once the occupation was over Greece remained crippled by civil war for several years, but the museum opened again under a new director, Christos Karouzos, and between 1945 and 1964 those artefacts that had not disappeared in the confusion were gradually retrieved and put back on display. The Antikythera mechanism survived it all, but by this time the excitement surrounding it was largely forgotten. With so much disagreement over its date and identity the science historians had moved on, while to the art experts and archaeologists now working at the museum the shabby-looking fragments could not possibly compare in importance to the beautiful vases and sculptures that filled the building's halls.

So the mysterious pieces were not put on display alongside the rest of the Antikythera haul. Once more they sat unnoticed at the bottom of a storeroom crate.

3

Treasures of War

The sight of him was made terrible by blasts of many trumpets and bugles, and by the cries and yells of the soldiery now let loose by him for plunder and slaughter, and rushing through the narrow streets with drawn swords. There was no counting of the slain, but their numbers are to this day determined only by the space that was covered with blood.

— PLUTARCH, *LIFE OF SULLA*

HER WOODEN STERN arched high out of the water, before swooping back on itself into a magnificent swan shape. Tucked under the hood of the stern was a tiny galley, its roof lined with terracotta tiles. Two men chatted inside as smoke from the cooking fire curled up past the swan's neck and into the starry sky.

Around them piles of clay jars with pointed bottoms were stacked against the wall, while beneath them two huge steering oars plunged into the racing water through openings high up in the tightly planked hull. Further along the deck the dim, yellow glow from an oil lamp marked the base of a sturdy mast, which rose from the centre of the ship and supported a striking square sail. A complex system of rigging allowed the sail to be raised and lowered by a set of evenly spaced vertical ropes, like a giant Austrian blind, giving it a scalloped appearance even as it bulged forwards in the wind. At the front of the ship the bow, like the stern, curved up and away from the waves. This gave the vessel a roundly triumphant air, while a smaller mast with a second square sail completed the effect by extending jauntily forwards.

She was moving unusually low in the water. Below deck, the hold was packed to the brim. Back at port the crew had used ropes and pulleys to lift the cargo off the quay and lower it through hatches into the hold. A series of massive statues had come first, made of the finest Parian marble – giant men and horses so heavy they strained the creaking ropes to their limit. Smaller bronze and marble figures, grand furniture, armour, glassware and finally a delicate clockwork mechanism were carefully packed into the remaining spaces and lashed into place.

Then she set sail, as the blank yet ever-open eyes of her cargo stared into the darkness of the hold, heading for a destination that she would never reach.

It's fair to say that up until the Second World War, scholars studying the Antikythera mechanism had floundered. They had discovered a few scattered references, including the month name 'Pachon' and a zodiac scale, which suggested that the device had an astronomical purpose rather than a navigational one and that it was therefore part of the cargo rather than a ship's instrument. Otherwise, despite plenty of speculation, it was still impossible to say anything certain about the mysterious fragments, beyond the fact that they were Greek and dated from somewhere between the second century BC and the third century AD.

Scrutinising the details of the gearwheels and inscriptions, however, wasn't the only way to investigate the mechanism. Once the wreck of the ship that carried it had been discovered, archaeologists also studied the rest of the salvaged cargo. Their discoveries help to paint a vivid picture of when the ship sailed, where her load was being taken and the sort of world from which she came. From there, we can guess at the origins of the Antikythera mechanism itself, and how it ended up on its final journey.

When the Antikythera wreck made headlines in the early 1900s, archaeology was taking its first tentative steps. Scholars knew of no other ancient ship to compare it with, so they had no way to judge very accurately who built it and when. The Archaeology Society's 1902 report on the Antikythera finds simply notes that the timbers of the hull, which seemed to have been held together by wooden tongues snuggled tightly into slots cut in the sides of neighbouring planks, indicated 'some curious method of boat building'.

Early on, it was the statues from the ship's cargo that got most of the attention. They were clearly recognisable as Greek and were dated to various periods between the fifth and the first centuries BC. Sparks flew in the debate about their exact age, but generally the bronzes were judged to be earlier and of higher quality than the marbles, which looked like later copies of classical originals.

The ancient Greeks are considered to have reached the height of artistic excellence in the fourth and fifth centuries BC. In the subsequent Hellenistic period (which runs from the death of Alexander the Great in 323 BC until the Romans took over in the first and second centuries BC), things went rather downhill. There was still much going on from an artistic point of view, but there wasn't the same innovation, the same magical inspiration. Sculptors and other artists, although of great technical skill, tended to look back rather than forwards and to copy the styles of the classical masters. In fact, despite apparent aberrations such as Archimedes and Hero, it has been assumed that the search for knowledge idled too. The great thinkers of ancient Greece still tend to be seen as those from earlier classical times, such as Socrates, Plato and Aristotle.

The other notable feature of the Hellenistic period is that the prominence of city states on the Greek mainland, such as the great old rivals Athens and Sparta, faded. They were rather too close for comfort to the growing shadow of Rome in the west. Young and ambitious Greeks headed steadily east into the huge and exciting empire that Alexander's conquests had opened up, particularly to Alexandria and Antioch, capitals of Egypt and Syria, which became the new centres of Hellenistic culture.

Valerios Staïs, director of the Athens National

Archaeological Museum when the fragments were discovered, plumped for the Hellenistic period, specifically the second century BC, as the most likely time when the Antikythera ship sank. As well as the statues, the sponge divers had brought up a collection of terracotta jars, called amphoras, which would have carried a range of supplies including the crew's food. Clams and oysters clung to them, but one of the sailors on the *Syros* managed to prise off some of the encrustations with his pocket knife and found both Greek and Latin letters denoting the jars' capacity. Staïs pointed out that it was during Hellenistic times, while the Romans were taking over but before they completely dominated, that Greek and Latin characters were used interchangably in this way.

However, this would have made impossible coin expert John Svoronos's dating of the inscriptions on the Antikythera mechanism to the third century AD and, sure enough, he had a different theory. Convinced that such a prestigious cargo of statues would not have gone unrecorded in ancient texts, he scoured the records of city after city for information. Eventually, from Argos on the Greek mainland, he found a document from the fourth century AD referring to artworks that he matched to those found in the Antikythera wreck. The mystery was solved, he announced proudly, in the newspapers and to anyone who would listen. The precious cargo had originated in Argos during the 300s, and was perhaps being taken east to Constantinople – by that time the capital of the Romans' eastern empire.

Svoronos's unorthodox methods aside, however, the statues couldn't reveal a precise date for the shipwreck. These classical works of art might have been preserved and displayed for centuries, just as the *Mona Lisa* or the Antikythera Youth himself are still admired today, so the pieces on the ship could have been hundreds of years old by the time it went down. But they did provide a crucial clue to the identity of the ship: the bronzes still had traces on their feet of the lead that had once attached them to their bases. These were not newly crafted figures being shipped for trade. They had already been on display for some time, before being wrenched from their pedestals in a hurry. In other words, the statues were stolen.

There was one obvious culprit: the Romans. From the middle of the second century BC onwards, their all-powerful armies spread out from Rome and gradually took control of the whole Mediterranean region. As they did so, they milked the conquered territories of their assets: slaves, food, gold – and art. The Romans loved foreign culture when they saw it and took as much of it as they could home with them to decorate Rome. A steady stream of statues, paintings, furniture, precious bronzes and silverware – both old classics and newly created works – arrived across the sea, filling the villas and palaces of emperors, officials and art collectors with masterpieces from Egypt, Asia and especially Greece.

Even Greek buildings weren't safe. Carved marble blocks and friezes were shipped back too; and used to decorate public buildings such as theatres and temples. The shipwreck at Mahdia in Tunisia, for example, which sank in around 100 BC on its way to Rome from Athens, was carrying a load of 70 giant marble columns.

From circumstantial evidence, then, one thing that Staïs and Svoronos could agree on was that the ship bearing the Antikythera mechanism was probably a Roman vessel, loaded with artworks and other treasures looted from Greek cities.

There the matter rested until many decades and two world wars later, when a new generation of archaeologists and divers became interested in the Antikythera wreck. In 1953 the diving pioneers Jacques Cousteau and Frédéric Dumas were travelling around the Mediterranean in their steamship *Calypso*, still looking for adventures that would challenge their recently developed scuba equipment. After hearing about Antikythera, and no doubt tantalised by the risks of the site, they decided to go and see if the sponge divers had left any treasures behind.

Dumas describes the expedition in his 1972 book, *30 Centuries under the Sea*. After being forced by one of the area's ubiquitous gales to shelter for a while at Kythera, the divers arrived at Antikythera's tiny port Potamos, where Kontos and his men had first hidden from the storm half a century before. This time Dumas was first in the water, which was so transparent he felt as if he might fall right down to the bottom. There was no sign of the wreck or its contents, but there was a dark, raised platform from which flowed the long green ribbons of posidonia plants. After exploring the area, Dumas was convinced that much of the ship was still there, buried in the seabed.

When a ship first sinks, its organic material becomes a veritable feast for sea creatures. Teredo worms (really a kind of mollusc), bacteria and other organisms eat their way mercilessly through the wooden hull, until it collapses under the weight of its contents. Eventually, any exposed wood disappears, leaving just a pile of cargo to mark the spot – like the heap of statues that had so stunned Elias Stadiatis on his first dive at Antikythera. But the entire hull doesn't necessarily disappear. As these creatures live and eat and die they produce particles of debris, which over time filter down to bury the lower parts of the wreck. This protects it by keeping out the dinner guests. As the centuries tick by the sediment gradually hardens from the bottom up, entombing the preserved remains within a case of limestone.

Subsequent dives, during which the divers dug with their hands like rabbits, turned up a few fragments of pottery, but not much else. Many stories had been told of riches that had rolled over the cliff below the wreck site, out of the sponge divers' reach – like the giant statues that the crew of the *Syros* had shifted in the belief that they were boulders – so one still evening Dumas ventured into the deeper water, guided down by *Calypso*'s anchor chain. The cliff dropped off to nearly 90 metres, at which point sand, punctuated by large blue gorgonian corals, stretched as far as he could see. Dumas didn't immediately find any remains from the wreck. But he couldn't risk stopping so deep down and his brain was thick with narcosis, so he worked his way back up the chain to the surface. He paid for his brief exploration over the cliff with only the second decompression injury of his long diving career – a shoulder so sore it felt as if he had fractured it.

Cousteau was unimpressed by the lack of pickings at the site. He spent the next few days exploring another ancient wreck that the party discovered just 300 metres away, with no statues but hundreds of amphoras – it must have been a trading ship, carrying staple supplies. But Dumas insisted on returning to Antikythera for one last dive before they moved on. He poked around with a crude suction device that he had made out of a metal pipe, powered by the air compressor in the boat above, and found various fragments of pottery as he pushed the pipe into the sand. Then, on his last attempt, he hit something hard:

> At the bottom of my last funnel, about two yards from the rocks, the pipe ran right into the hull of the sunken ship, which was perfectly preserved under a foot and a half of sand. If only I had more time!

He could even see traces of paint on the 2,000-year-old wood. The lower parts of the ship, along with any cargo buried as the sediment settled, were surely still fresh and intact under the sand . . . Bound by the laws of nitrogen, Dumas had to surface. But as *Calypso* sailed the next day towards other adventures, his head was full of plans to return and dig further. He had no idea that it would be more than 20 years before those plans came to fruition.

The next advance towards understanding the origins of the wreck came in Athens. As Dumas and Cousteau sailed from Antikythera, an American archaeologist called Virginia Grace was working with a young Greek graduate, Maria Savvatianou, on a project to catalogue 25,000 broken amphora handles that had been found at sites across the Mediterranean, but were now lying uncatalogued at the National Archaeological Museum.

Savvatianou was searching for clues to help identify where the different amphoras had come from, and in 1954 she came across Svoronos's publication from 1903, in which he described the finds from Antikythera. It included a black-and-white photo showing some of the salvaged amphoras in a terracotta line-up against the wall. An excited Savvatianou showed it to Grace. Using the photo as a guide, it might be possible to identify the Antikythera amphoras from the unlabelled piles at the museum.

When the jars were first salvaged, they had been of limited interest to archaeologists. Back then there were no other wrecks around, and although ancient Greek and Roman artefacts had been found at various land sites around the Mediterranean, few had been accurately dated. All scholars could do was give a rough estimate of the century from which an object might have come. But archaeology had changed a lot since 1900. By the 1950s hundreds of wrecks had been discovered (the discoveries have continued – there are now well over a thousand sunken ships from before 1500 AD known in the Mediterranean alone). And the study of land sites across the region had become a much more precise art, with the squashed layers of ancient remains peeled apart and correlated to particular historical events.

If Grace could find the Antikythera amphoras, she had a good chance of dating them and therefore the wrecked ship. But as she looked at the rest of the photos in Svoronos' paper, she realised that other uncatalogued artefacts, including pottery and glassware, should be identifiable too. So Grace enrolled some friends and colleagues from the intimate community of American archaeologists working in Athens, with the intent of finding out when the ship sank and where it sailed from. Henry Robinson was the kindly director of the American School of Classical Studies at Athens (ASCSA) and an expert in early Roman pottery, while his friend Roger Edwards, who was based at the University of Pennsylvania in Philadelphia, was a specialist in Greek Hellenistic pottery and a frequent visitor to Athens's ancient marketplace. Gladys Weinberg of the University of Missouri would check out the glassware, and the journalist and archaeologist Peter Throckmorton would study the shrivelled fragments of wood from the boat itself.

Getting access to the museum storerooms was a masterclass in cajolery – the Greek staff were touchy to say the least about allowing a team of foreign experts access to the museum stores. And even once the team was in, their task was far from straightforward. Like the amphoras, most of the objects from the wreck had never been properly catalogued or labelled, and the chaos caused by the Second World War, as well as changes in the museum's management, meant that a significant proportion of the finds described in 1903 had been lost altogether. Working out which of the dusty items in the stores came from Antikythera meant comparing them against the few that had been drawn or photographed at the time, then looking around to see what else seemed to be part of the same batch.

Throckmorton was an experienced diver who had worked with sponge divers to explore several ancient wreck sites in the Mediterranean. Grace knew him well, because she had often helped him to identify the amphoras he recovered from wrecks. One of his expeditions was the ingot-filled galley at Cape Gelidonya, off the sunbaked coast of Turkey, which he had excavated with a team that included Frédéric Dumas. The Frenchman didn't like the wealthy American much, complaining that he was more interested in collecting pricey souvenirs for himself than in the scientific value of the wrecks he explored. At Cape Gelidonya, Dumas says that Throckmorton 'started chipping algae off the first raised block to expose the ingots before we even got it to the beach. I had to ask him to stop and follow scientific protocol'.

Nevertheless, Throckmorton's work in the Mediterranean was hugely influential at a time when marine archaeology was just starting to become a scientific subject (it was always a few steps behind its terrestrial cousin). And few people knew more than he did about the art of ancient shipbuilding.

In the museum stores Throckmorton found a dozen pieces of desiccated planking lying in a crate, shrunk to a fraction of their original size as the water that had propped up the cells in the wood for so long dried out. The wood was elm, from central Italy, indicating that the ship had indeed been built by the Romans (the Greeks tended to use Aleppo pine) And the strange construction that had so intrigued the archaeologists in 1901 was now instantly recognisable to Throckmorton as a 'mortise-and-tenon' or 'shell-first' construction.

Modern wooden ships tend to be built frame first, the planks forming the sides being fitted after the skeleton frame has set the shape of the hull. But the ancients built their boats the other way around. The planks were tightly slotted together using wooden tongues called tenons, in a precise manner that Cousteau once described as 'more akin to cabinet-making than ship carpentry'. Only after the planks had been fixed together to form a smooth, watertight hull were the ribs of the frame attached inside.

This method of construction lasted for more than 3,000 years, from the Egyptians to the Romans, despite being incredibly labour intensive. It's not clear why it took so long for shipwrights to switch to the much cheaper and easier frame-first method; maybe because mortis-and-tenon ships were so well built that they literally lasted for centuries.

Throckmorton was frustrated that the museum staff wouldn't allow him to remove any of the wood from Athens for further study. But one tiny fragment was eventually sent away for a test that would have been unimaginable for Svoronos and Staïs: carbon dating.

All carbon atoms have a nucleus that contains six protons – that's what defines them as carbon. The number of neutrons in the nucleus can vary, so you can get carbon-12 (which has six neutrons) or carbon-13 (seven neutrons). Both are stable, and will stick around in the atmosphere for as long as you like. But you can also get tiny amounts – just one part per trillion of the carbon on the Earth – of carbon-14 (eight neutrons). It is made when energetic particles from space called cosmic rays strike atoms in the upper atmosphere, causing them to spit out neutrons. If one of those neutrons hits a nitrogen-14 atom (seven protons, seven neutrons), it jumps into the nucleus and kicks out a proton for good measure, resulting in carbon-14. This form of carbon is unstable, however, and it undergoes very slow radioactive decay back into nitrogen-14.

This carbon-14 filters down into the air around us and as we breathe, we constantly take in tiny amounts of it, and incorporate it into the molecules of protein, carbohydrate and fat that make up our bodies. It gradually reverts to nitrogen, but as long as we stay alive we keep breathing in more of it, so the level of carbon-14 stays roughly the same as that in the atmosphere. Once we die, however, that process stops. The amount of carbon-14 in a dead body then falls very slowly as the individual atoms decay, halving for every 6,000 years or so that passes. The same holds true for any living organism, from a sailor on a ship, to the ship worm eating holes in the wooden hull beneath him, to the trees that first made the hull's timbers.

By measuring the amount of carbon-14 in a piece of wood and comparing it to the amount of carbon-12 and carbon-13, which stays constant, it's possible to work out roughly how long it has been since the tree was cut down. The test could give the first scientific indication of the date of the Antikythera wreck.

The fragments were sent to Elizabeth Ralph, one of the only scientists in the world with the expertise to carry out the new-fangled technique. She had worked with its inventor, Willard 'Wild Bill' Libby at the University of Chicago, but since 1951 she had been working at the University of Pennsylvania, where she set up the country's second carbon-dating lab in the basement of the physics building.

Ralph's results soon came back: she concluded that the wood dated from between 260 and 180 BC. That gave 260 BC as the earliest possible limit for the wreck. But the ship could have sailed a fair bit more recently than 180 BC, she pointed out cautiously. Although the cells in our bodies are recycled throughout their lives, trees work differently. Once the rings in a tree trunk form, they are effectively dead – the life and growth of the tree happens around the edges as the trunk thickens, and at the tips of its branches and the ends of its roots. If the bit of wood that was tested came from the middle of the trunk of a very large elm it would have been laid down when the tree was a young sapling.

The mature tree wouldn't have been cut down until many decades later. The wood itself could also have been quite old when the ship was built, and the ship could have been quite old when it sank.

Still, the result was consistent with Staïs's idea that the ship sank in the second century BC, and effectively ruled out Svoronos's theory that it sailed from Argos some 500 years later.

The shrivelled pieces of wood also allowed Throckmorton to estimate the original size of the planks used, from which he concluded that the ship was a heavily built merchant ship. It was perhaps 30 or 40 metres long and able to carry a hefty 300 tons of cargo, making it one of the larger vessels to sail the ancient Mediterranean. The characteristic arching sterns and gathered square sails of these ships are pictured on several manuscripts, pottery fragments and mosaics.

The objects the ship was carrying told his colleagues even more. As the original Greek archaeologists realised, dating the statues doesn't help much in dating the ship. So the rest of the team focused on the more mundane items on board, largely ignored until that point by everyone except for Staïs – the pots, plates and jars that would have been part of everyday life for the crew. These weren't as glamorous as the gleaming artworks, but such cheap, easily breakable items wouldn't have lasted long on rough ocean crossings, and were therefore unlikely to have been more than a few years old when the ship sank. Amphoras carrying the ship's provisions would probably have been picked up at the ship's most recent ports of call, so might also give a clue to its final route.

By the 1950s there was plenty of smashed pottery from other digs for the experts to compare with the Antikythera finds. Much of it had been unearthed over the previous two decades in Athens' famous Agora or marketplace. Athens was a centre of trade for the whole region, so goods and containers from across the Mediterranean found their way to the hubbub of its central market. The most helpful remains for the archaeologist are those from great disasters or celebrations – past events involving such dramatic destruction or growth that the buried remnants form recognisable layers in the ground. Correlating these layers to dated written records provides exact time markers against which to assess new finds.

One ancient event that's squarely in the disaster category is when the great Roman general Cornelius Sulla smashed the city of Athens in 86 BC in the last years of the Roman republic. The pieces lay trampled beneath the feet of generations of Athenians until archaeologists such as Virginia Grace and Gladys Weinberg carefully brushed away the earth, revealing the horror and hunger that had been frozen there for 2,000 years.

The Romans had extended their grip east as far as the Asia Minor coast and Cornelius Sulla was on his way to deal with King Mithridates of Pontus, which is just south of the Black Sea. Mithridates had ambitions of greatness. After killing several of his brothers (and marrying his sister) to clear his way to the throne, he was intent on expanding his kingdom. Two years earlier, Mithridates had led the Greek cities in the region in a huge rebellion against Rome, involving the synchronised bloody slaughter of some 80,000 Roman citizens.

Now Sulla was on the warpath. With fiery hair and gleaming grey eyes he looked fearsome, and that fire burned just as strongly within him. He stopped off at Athens on the way to Asia Minor, because Mithridates had installed a puppet leader there called Aristion. Sulla laid siege to Athens for months, during which the starving citizens were reduced to eating grass and shoe leather, but Aristion danced on the ramparts. He and his jesters taunted Sulla, laughing at the general's blotchy complexion and casting aspersions on the character of his wife Metella.

When his army finally made it into the city late one night through a poorly defended section of wall, Sulla was in no mood to be merciful. Driven on in the moonlight by their leader's rage, the soldiers killed the cowering Athenians and left the shining city in pieces. As the Greek biographer Plutarch later observed, the corpses were too many to count, the number of dead only being estimated from the amount of blood that flowed through the streets and out of the city gates.

The screams soon faded. But the dense layer of debris that archaeologists later uncovered in the Agora has become a key reference point for dating artefacts from the first century BC – not just from Athens but from Rome and the rest of the Hellenistic world.

Another easy-to-date layer in the Agora comes from the 10s and 20s BC. It consists of the remains from the vigorous burst of activity and construction in the city in the early years of the reign of Augustus Caesar, Rome's first emperor. This was a new order: the style of the pots and jars is markedly different from those in the layer testifying to Sulla's night of destruction. By comparing the two (and other dated layers in the Agora and elsewhere) archaeologists can work out how the designs of various jugs and jars changed over the years, and slot any new finds into that timeline.

Back in the Athens storeroom, pottery experts Roger Edwards and Henry Robinson investigated the Greek and Roman crockery from the Antikythera wreck. They agreed that both dated from soon after the Sulla layer, perhaps between 86 and around 50 BC. Surprisingly, none of the pots were from Athens itself, one of the obvious places that the Antikythera ship could have sailed from. Instead, they came from exotic eastern cities on the Asia Minor coast (in what is now Turkey), including wine jugs from Pitane, Pergamon and Chios, and an Aladdin-style oil lamp from Ephesus.

Meanwhile, their colleague Gladys Weinberg studied the glass plates and bowls. Weinberg was a striking, athletic woman who had worked in the secret service and as a journalist, but her excavations at Athens and Corinth had also made her an expert in ancient glassware. She saw immediately that the glass items from the wreck were very different from the pottery – they were luxury pieces of the highest quality and beautifully preserved. Rather than having been used by the crew, they were probably part of the cargo. 'Looking at them, in their pristine, almost flawless condition,' she wrote, 'one thinks it impossible that they could have been found on the sea bottom, and how they survived seems a mystery.' Like the Antikythera mechanism, each of the glasses had become covered in a hard crust of limestone that protected them while they were under the sea, but was subsequently removed by museum staff. Such a coating damages the surface of pottery and of marble statues, but it can't find a grip on smoothly polished glass and can be easily chipped off to reveal the original beauty beneath.

Among the finds was an elegant blue-green bowl carved with an understated floral design that would grace the most exclusive of today's shop windows, as well as a set of mosaic dishes in which stripes of rose, purple, green, yellow and aquamarine glass have been coiled into tiny spirals and melted together with breathtaking attention to detail. Many of the pieces were unique or the earliest of their kind to be found, so in this case the flow of knowledge was reversed: the dating of the wreck became an important reference point for setting the chronology of other similar glass vessels.

The most precise information about the wreck, however, came from Virginia Grace, who squeezed her answers out of the ship's amphoras. Rounded vessels with handles at the top, amphoras have a narrow neck that can be stoppered and a pointed base that serves as a grip when pouring. They were used for all sorts of foodstuff, such as grain, olives, wine, even pickled fish, and were everywhere in the ancient Mediterranean – archaeologists have found hundreds of thousands of them. Private homes and shops had dedicated stands for them, while in warehouses they were leant against the wall or dug into the sand, and on ships they were stacked tightly against the sides of the hold, often several layers high.

The archaeologists overseeing the original salvage operation didn't know what to make of the jars when the sponge divers first pulled them up to the surface and deposited them, dripping, on to the deck of the *Syros*. They were so bemused about the presence of so many different kinds of amphoras alongside the precious statues that the ship's captain teased them by suggesting that the owner of the Antikythera ship, clearly not satisified with raiding temples and agoras, must also have looted pottery markets and grocers' stores in every port he visited.

But the tall, delicate 'Miss Grace' was a world authority on the subject. Except for when she fled Athens for Cyprus during the Second World War, she worked in a huge marble and limestone building that had been reconstructed, columns and all, in the Agora. She had studied thousands of amphoras and her basement was full of them, of all sizes and shapes, painstakingly glued together from fragments. In Grace's careful hands each reconstructed jar revealed a story it had been waiting to tell for millennia. The length of a neck, the slope of a shoulder, the sag of a belly . . . these subtle differences in design spoke volumes to her of the place and time from which each individual piece originated.

An angular profile and round handles identified most of the Antikythera jars as coming from Rhodes, while a slim body and slightly careless manufacture suggested they were made in the first century BC, just after Sulla's rampage in Athens. Earlier amphoras were more precisely made, whereas Rhodian jars from the new order Augustan deposit at Athens were even shoddier (production standards on Rhodes understandably slipped somewhat after the Roman general Cassius sacked the island in 43 BC).

A few of the Antikythera jars had double-barrelled handles (with a figure-of-eight cross-section, as in a double-barrelled shotgun), suggesting they were made in Kos – probably filled originally with Koan wine, the region's best. Grace dated them against debris found on the nearby island of Delos, with bloody disasters again providing the necessary timeline. The style of the Koan jars from the wreck resembled that between fragments trampled to the ground when Mithridates' soldiers attacked Delos during the rebellion against the Romans in 88 BC, and remnants of a pirate raid that destroyed what was left of the luckless island in 69 BC.

None of the experts found any objects from Athens, or from anywhere else on the Greek mainland for that matter – everything originated from Asia Minor or the eastern Aegean islands of Kos and Rhodes. In a joint paper finally published in 1965 they concluded that the ship must have started her journey somewhere on the Asia Minor coast between 86 and around 60 BC. Once loaded to the brim with statues she must have sailed west towards Rome, stopping at the trading ports of Rhodes and perhaps Kos to pick up essential supplies on the way.

The sturdy vessel then headed west past Crete and through the channel between Crete and Cape Malea, where the jagged rocks of Antikythera cut short her journey. Otherwise the captain's plan would have been to hug the coast up the western coast of mainland Greece and hop across the Adriatic Sea to Brindisi or Tarentum on the heel of Italy. Then he would have sailed her south around the toe, either all the way around Sicily or through the Messina strait, before the final run north up the west coast of Italy towards Rome.

It has been suggested that her cargo belonged to Sulla himself; goods looted from the Asia Minor cities after his war against King Mithridates. No stranger to confrontation at home as well as abroad, Sulla marched his armies through Rome twice, then briefly enjoyed the absolute power of a dictator, before he unexpectedly resigned to spend what remained of his old age partying among the actors, musicians and dancers that he loved. Though he willingly relinquished his dictatorship after a couple of years, he set an ominous precedent in the transformation of Rome from a republic to an empire, as this was the first time the Republic had been led by one man.

After forcing a weakened Mithridates back to his homeland in 84 BC, Sulla had cut a devastating trail through Asia Minor. He ravaged the cities there with infamous bloodthirstiness and greed, then shipped the booty home, among other things to pay for a huge triumphal parade that helped to cement his popularity with the people of Rome. The Antikythera ship might have belonged to him, or to one of the generals later installed to rule the territories there and extract the crippling taxes that paid for Rome's excesses.

The Syrian satirist Lucian, who travelled around the Roman Empire in the second century AD, even wrote about a particular ship full of Sulla's loot that sank off Cape Malea. He mentioned it in a story he told about a painting by the famed Greek artist Zeuxis (a striking fantasy that depicted a female centaur nursing a pair of infant centaur twins), because the ship was supposedly carrying this artwork when it sank. It's likely, however, that this ship would have sailed from Athens, because Lucian says that is where a copy of the painting was held. So, although commentators have sometimes tried to link the two, it probably wasn't the wreck found at Antikythera.

Although it is tempting to try to identify the Antikythera wreck with a particular ship described in the written accounts that have survived through the centuries, the chances of a match are tiny. Countless ships were lost in the area at around this time, and the vast majority were never recorded and will always remain nameless. Historians estimate that what with the treacherous waters along this route as well as pirates (who infested the sea around Antikythera) and the greedy overloading of ships, as many as 5 per cent of the booty extracted from the eastern Mediterranean ended up at the bottom of the ocean. We know from more recent experience that even in the most favourable conditions carrying such heavy cargo is hazardous. When Lord Elgin took the Parthenon marbles to London in 1803, one of his ships was wrecked off the coast of Kythera. And a French steamer carrying a bronze statue of Ferdinand de Lesseps – the diplomat responsible for the construction of the Suez Canal – to Egypt for the canal's grand opening in 1869 almost sank when the massive figure shifted in the hold.

We can still get a little closer to the last days of the Antikythera ship, however, thanks to Frédéric Dumas and Jacques Cousteau, who finally returned to the site of the wreck in 1976. By this time Cousteau's books and films about the wonders of the sea had made him famous throughout the world. They were back with their ship *Calypso* in the Mediterranean, working with the authorities in Greece to visit a number of ancient wreck sites, and Cousteau wanted to make a film about Antikythera. The two Frenchmen were confident from their previous visit that the wreck still contained secrets for the taking.

They stationed *Calypso* immediately above the wreck. Two anchors were dropped on her seaward side and three heavy nylon lines looped over rocks on the shore. The hope was that in this treacherous location the web of ropes would hold the ship tightly in place and save her from being bashed against the rocks if a storm suddenly rose. It was a long way from the trials of Kontos and his men. This team of divers had the latest scuba gear, with sleek black and yellow wetsuits and airtanks in matching cases, not to mention suction-powered digging equipment and powerful floodlights to aid their work. But, as ever, they had to be careful of the bends. Each diver could only dive twice a day, with 20 minutes on the bottom each time. He paid for this pleasure with a lonely half-hour decompression stop on the way back up, clinging to the bottom of *Calypso*'s keel three metres below the surface.

Most of the time that didn't include Cousteau himself. The film shows him proudly emerging from the water, but staff at the Athens museum now scoff that he didn't actually dive, but 'only turned up for the cameras'. Maybe that's fair enough, as Cousteau and Dumas were in their sixties by then. Dumas was always the dreamier of the two, noticing how the sunlight scattered on the water, or making friends with the octopuses who made their homes inside sunken amphoras. The force of nature that was Cousteau was much more interested in the PR side of things, at which he was extremely accomplished. He referred to all his documentaries, including the Antikythera film, as 'advertising'. What he was advertising – the sea, the wreck, himself – wasn't clear. Perhaps it didn't really matter.

After meticulously photographing and mapping every square metre of the wreck site, the team of divers set about digging through each section in turn. It was clearly a much more scientific survey than the original salvage expedition in 1900. The divers recorded the position of everything they found before placing it in a basket or tying it to a rope to be raised to the surface. However, some of the techniques they used might cause today's archaeologists to gulp. The team's prized weapon was a more powerful version of the suction pipe that Dumas had used in 1953. Like a souped-up vacuum cleaner, it devoured whatever came into its path – water, silt, artefacts – and sucked it up to the surface, where it was spewed out into a basket strainer that hung off the side of the boat. Every so often the men on deck (as in 1900, there were no women on board) would sift through the basket looking for fragments of value – often in rather smaller pieces than when they had been dragged from their resting places on the seabed. This time, though, the archaeologists wore skimpy swimming trunks rather than smart suits.

Progress was slow, and the team didn't unearth any of the big statues they had hoped for – perhaps the sponge fishers really did get them all, or they were buried deeper than the vacuum pipe could suck, or perhaps the forming sedimentary rock had already engulfed them. The divers were also on the lookout for any missing pieces of the Antikythera mechanism, but had no luck there either. Still, the expedition was rewarded with an array of evocative objects, including another oil lamp, a pristine marble finger and thumb, and a rather more battered hand and foot. Then came an ornate gold cap that once served as the setting for a precious stone, some giant bronze ship nails, and the magnificent Spartan-style crest of a bronze helmet. Cousteau was especially pleased with two bronze statuettes on rotating bases. One was a boxer, his strong right arm stretched forwards, the other a rather camp young man with hands raised above his elbows and hips provocatively swayed. Then came the most chilling find so far. A human skull.

It's actually quite rare to find human remains in wrecks – sailors tend to struggle and swim before they drown, so their bodies get carried off in the current, then eaten or swept up on shores far away from the site of the original disaster. This unfortunate must have been trapped inside the ship as it went down – a crewman snoozing after a night on the Koan wine perhaps, or a captured pirate confined to the hold.

The bits and pieces were all taken back to the Athens museum and recorded on Cousteau's film, but like the sponge divers' haul they were never properly catalogued or displayed and the finds have never been formally published.

Except for one. A misshapen lump of silver metal, small enough to fit comfortably in your hand, confused its discoverers at first. It turned out to be the most significant find of the expedition: a stack of coins, fused together by the chemical action of the sea in the shape of their original container, which had long since rotted away.

Coins are an archaeologists' dream when it comes to dating sites, because they carry markings to identify who issued them and they don't tend to stay in circulation for very long. Svoronos would have been blown away by this find. Once the pieces in the lump had been cleaned and separated, they turned out to be silver coins from the city of Pergamon, worth four drachmas each. One, bearing the image of an ivy-wrapped basket of sacred snakes, carried the initials of a magistrate who had served in the city from 85 to 76 BC.

There were several bronze coins in the same hoard – they were in a sorry state, but two of them could be identified as coming from the city of Ephesus, which is about 100 miles south of Pergamon. Artemis, goddess of the hunt, stared out from the front face of each coin with bow and quiver at her shoulder, while revealed on the back was a kneeling stag and the inscription *Demetrios*, perhaps the name of the issuing magistrate. They were slightly younger than the silver pieces, issued in 70–60 BC.

These coins pinned down the ship's origin even more precisely than Grace's reading of the amphoras' curves. It must have sunk sometime between 70 and 60 BC, and it had probably set sail from Pergamon on the Asia Minor coast, where most of the coins had been minted.

This makes the ship a little late to have carried Sulla's loot – after a life spent partying as hard as he fought, Sulla died from liver failure in 78 BC. But by this time a new young general was ravaging the eastern Greek cities – and fighting the persistent King Mithridates, who was still managing to drive the Romans to distraction. The name of this upstart was Pompey the Great. He travelled to Asia Minor and finally defeated Mithridates' troops in 65 BC, though even then he still didn't manage to kill him. A fleeing Mithridates tried to take his own life by poison, but years of consuming a small dose of poison every day to strengthen his constitution had rendered him immune to it; in the end, it is said, a servant had to run him through with a sword.

Pompey had long clashed with Sulla in Rome and if anything he was more greedy than his elder, and more desperate to prove his glory back home. With boyish features and self-consciously swept back hair, his supporters said he looked like Alexander the Great, although Sulla is rumoured to have given him the nickname 'Great' as a sarcastic joke. Pompey had a talent on the battlefield (and on the sea – it took him just three months in 67 BC to clear the Mediterranean of pirates) and he was efficient at managing provinces once they were conquered, and ensuring a constant flow of goods and treasures back to Rome. After beating Mithridates, Pompey killed and looted his way through Pontus, Syria, Palestine and even Jerusalem, turning them all into Roman territories.

When he returned to Rome in 61 BC he held the greatest triumphal parade in the city's history, so huge that the spoils were carried into the harbour on 700 ships and it took two days for the spectacle to pass through the streets. Placards inscribed with the names of the many lands he had conquered were followed by his troops, royal prisoners, spoils from the raided cities, exotic animals from his travels, and gold and silver statues of his dead enemies, including Mithridates. It didn't all go as Pompey had hoped, though. According to the ancient historian Plutarch he had planned to enter the city on a gem-studded chariot drawn by four elephants that he'd brought back from Africa. Unfortunately, they wouldn't fit through the city gates and he had to switch to horses at the last minute.

It was a sign of things to come. By the time Pompey got back to Rome another new star had risen: Julius Caesar (the great uncle and guardian of the boy who was to become Emperor Augustus). Caesar was a gifted general, too, and much cleverer politically than Pompey. Both hoped to become Rome's next dictator, but after several years of manoeuvring the prize was awarded to Caesar. Pompey was unceremoniously stabbed in the back when he fled to Alexandria in 48 BC, by officials who hoped to impress the new Roman leader.

We may never know for sure whose cargo was on the Antikythera ship, but the latest dating of the wreck fits the time that Pompey's troops were sending back bounty from his eastern conquests – perhaps the statues were reparations extracted from Pergamon and Ephesus after the wars with Mithridates. With the rich nature of some of the cargo, in particular the bronze statues, gold jewellery and ornate furniture (the marble statues are still thought to be newer copies of classical originals), maybe some of the items were even destined to be carried through the streets of Rome in the triumphal parade.

So was the Antikythera mechanism also taken from Asia Minor, or could the ship have picked it up at a subsequent port of call? Pergamon, where the ship most likely set off from, was an extremely rich and civilised city, and a scientist working there at the time would certainly have had access to bronze and to engineering skills. But from the scattered records we have access to today, we don't know of any major astronomers or instrument-makers based in Pergamon at the time. And by the first century BC the Romans had taken over the city, so scientific activity may have been on the wane.

At Alexandria the royal family, the Ptolemies, paid for a huge research institute attached to the city's famous library, the Museum of Alexandria, where many ancient scholars would work at some time or another. The Antikythera ship might have stopped off at Alexandria, but the evidence is circumstantial – it's not too far out of the way, and some of the luxury glassware it was carrying could have been made there, but similar items have since been found in several other sites around the Mediterranean. And by the 60s BC the importance of the school was at a low. Academic activity had been interrupted by King Ptolemy VIII, who persecuted and expelled the city's intelligentsia in 145–144 BC and put the library under the control of the army. It was many decades before any scholars of note appeared again.

Rhodes, on the other hand, seems a more promising prospect. Because of its location in the south-east Aegean, the island was a key trading centre, especially for grain from Egypt. Virtually every passing vessel would have stopped there for supplies, and the large number of Rhodian amphoras found at the wreck site suggests that the Antikythera ship was no exception. Rhodes was extremely wealthy because of its strong trading position, and its capital city glittered with thousands of bronze and marble statues. 'Rhodes had more statues than trees,' staff at the Athens museum still like to say. 'And it had a lot of trees . . .'

The Rhodians tried hard to stay politically neutral as the Romans clashed with all around them, and, helped by their impressive naval fleet and siege defences, they managed to stay relatively independent as the rest of the Aegean succumbed, at least until the island was brutally sacked in 43 BC. As the second century BC turned into the first, it was one of the few cities where scientists were still free to work, and we know that there were some big names there, particularly in the field of astronomy. Two notable examples are Hipparchus, the possible inventor of the astrolabe, who lived on Rhodes in the second century BC, and a teacher called Geminus, who made observations there several decades later.

The islanders were on reasonably good terms with Pompey, who visited Rhodes several times. They would have engaged in trade with his ships as they passed. But they would also have offered treasured possessions as gifts to ensure Pompey's protection, and the general or his men would have felt free to carry off any valuable or intriguing items that caught their fancy. A mechanical computer! How better to impress the notables back home.

But what was this computer? How did it work, and what was it for? While Virginia Grace and her colleagues scrutinised the salvaged pots and plates, an English scientist called Derek de Solla Price began to decode the device itself. What had ultimately thwarted all those who had studied the Antikythera mechanism before him was the fact that their research was limited to the barely legible details on the surface of the broken pieces. Price used the developing technology of X-rays to see what was hidden beneath.

4

Rewriting History

Knowledge works rather like a large jigsaw puzzle. You wait until somebody puts down a piece and try to find a piece of your own to place on that living edge.

— DEREK DE SOLLA PRICE

DEREK DE SOLLA PRICE fell in love with many things. One of them was Athens, the first time he saw it in the summer of 1958. It is dirty and loud, and uncomfortably hot, and compared to his home city of London the atmosphere is abrupt, almost angry. But a few minutes walk from the car fumes and hooting of the busy Constitution Square, the winding back streets of the Plaka district delight him, with their little shops selling brass and coffee and spices and flowers.

Every so often the cobbled street opens out into a tiny square, and squashed between the other buildings a little Byzantine church or Ottoman mosque appears – the various mosaics, spires and arches telling the story of the city's colourful past. Then, as Price keeps walking south, the alleys open out in front of him and suddenly the noise and the centuries fall away. Straight ahead, the steep rock of the Acropolis draws his eyes upward and from the top rise the perfectly symmetrical lines of the Parthenon temple, still as awesome as when the Athenians dedicated it to their virgin goddess in the fifth century BC.

At the foot of the hill lie the remains of a market place. This is the Roman Agora (the older square of classical Greek times is a little further away, with no visible trace above ground until archaeologists started their work there a couple of decades before). Here, in the quiet of early morning, dozens of broken columns rest on the dry grass, with lonely heads, bodies and legs from once-proud statues strewn among them. Just one building – an octagonal marble tower decorated with carvings representing the eight winds – stands intact as a reminder of the gleaming city centre this once was.

But it isn't the Parthenon or the Tower of the Winds that Price is here to see. Before the glaring sun rises too high, he heads the 20 minutes or so walk across town to the elegant rectangular building that is the National Archaeological Museum. Inside, in a basement storeroom, he is finally going to see the mysterious object that has brought him all this way.

Price is 33. In a sense, his whole career so far has been a preparation for this moment. All of his various passions and interests have converged in leading him to this single artefact. It holds, he hopes, the ultimate answer to the questions he has been asking all his working life.

Some of the questions started even earlier, during his childhood in London's East End. He was born in 1922 to Philip Price, a tailor, and Fanny de Solla, a singer. The couple didn't have many material possessions, but they had enough money to indulge their young son in his love of Meccano, which was all the rage at the time. With enough ingenuity, the red and green painted girders, pulleys and cogs could be built into pretty much anything a boy could imagine – a bridge, a crane, a car, a spaceship – and Price wasn't short of either ingenuity or imagination. The toy instilled in him a passion for mechanics and for how things work, which stayed with him for life. One of his favourite stories when he was older was about the Scottish physicist James Maxwell, who, growing up in Edinburgh nearly a century earlier, once asked a workman operating a piece of machinery, 'What's the go of it?' Frustrated to receive only a vague reply, the boy stamped his foot impatiently. 'No! No! What's the *particular* go of it?'

Price recognised this urgent desire to know in his own character, although the ease with which he compared himself to one of the country's greatest scientists is perhaps equally revealing. When he wasn't building models, the young Price snapped up science-fiction pamphlets, printed on cheap paper and decorated with sensational cover images, with names like *Amazing Stories* and *Marvel Tales* and bold stories inside that took him far from the grey world of 1930s London. At school, he showed an aptitude for physics and maths, and plucked up the courage to send some of his original proofs to the eminent Cambridge mathematician G. H. Hardy.

Despite his talent, Price didn't have the money or the background to go to university, so he followed a less conventional route to pursuing the subjects that he loved. He got a job as a lab assistant at the newly opened South West Essex Technical College, which enabled him to study part-time for a degree at the University of London. The physics equipment there was one glorious step up from Meccano. Square and black with clunky dials and flickering green screens, the oscilloscopes, voltmeters and spectrometers were as heavy as stones, and packed full to bursting with valves and wiring. With such instruments you could make sense of things; you could measure the whole world! Price spent hours taking these devices apart, tinkering with them and putting them back together, until his fingers and his heart were intimately familiar with their workings.

He got his degree in physics and maths in 1942 and the college – seriously short-staffed because of the war – instantly promoted him to lecturer. He worked in one classroom, often for eight hours straight, as different sets of students passed through, learning the curriculum as he taught it. He also carried out research for the military on the optics of molten metals, and the University of London awarded him a PhD for it in 1946. Once the war ended, however, there was no job for him in London, so he took two big leaps into the unknown. He accepted a teaching position at the young Raffles College in Singapore. And he married a Danish girl called Ellen Hjorth.

Singapore was wonderful and exotic and it inspired in Price a new love for oriental culture and its history. It also introduced him to the history of science. Raffles College acquired a full set of the *Philosophical Transactions of the Royal Society* – the journal of Britain's foremost scientific body, with such worthy members over the centuries as Humphry Davy, Isaac Newton and Robert Hooke. The college library was still being built, so Price seized his chance and took the beautiful calf-bound volumes home with him – into 'protective custody', he joked. Accustomed by now to teaching himself everything, he used them as bedtime reading, starting with the first volume from 1665 and working his way through. In the journal's pages he learned how scientific knowledge had gradually accumulated, with each generation of scientists building on the work of the next to bring the world around them into ever sharper focus.

As he read, Price put the finished volumes into neat chronological piles on his bedside shelves. Then he noticed something strange. Though all the stacks covered the same number of decades, each was twice as tall as the one before. He stared at the pattern, trying to absorb what it meant – curves and lines and numbers flashing through his mind faster than his linear, more logical thoughts could keep up. He had always thrown himself into physics because it was measurable, dependable. It turned an uncertain world into numbers that followed rules. And once you knew the rules of the world you could understand it, predict it, control it – from the bouncing of a billiard ball to the splitting of an atom. Physics had its limits, of course. He had learned to accept that. It couldn't help you understand history or knowledge, for example, any more than truth or love. How could you plot knowledge on a graph?

But here, against his bedroom wall, it had happened. The scientific knowledge acquired over centuries had stacked itself up on his shelves and was displaying itself to him as a beautiful exponential curve. In other words, it showed a perfect and predictable mathematical relationship that doubled and doubled and doubled over time, as regular as clockwork. Of course! By counting scientific papers, you could measure science as it advanced. Price rushed to the university to check every other journal he could find, adding them together for each subject then piling up the volumes with trembling hands. For every one it was the same – the size of the stacks followed the same pattern. From Isaac Newton laying the groundwork of classical mechanics to Ernest Rutherford probing the atomic nucleus in the twentieth century, it made no difference – the pages inscribed with the scientists' results fitted the curve; they grew exponentially over time. Price had discovered the law that governed the path of knowledge itself.

That moment set him in a new direction. He felt his insight opened a window to a bright and certain future in which scientists would illuminate the uncertainty of the world at an explosive rate, until there could be no dark corners left. Price discussed this discovery excitedly with his badminton partner, a young British historian called Cyril Northcote Parkinson; the two men batting ideas back and forth breathlessly as they volleyed the shuttlecock over the net.

Not to be outdone, Parkinson formulated his own law from these verbal matches, which he felt was just as revolutionary. The inexorable growth of bureaucracies, too, could be described by a mathematical equation. Not bad, said Price, but he bet Parkinson that his own law would bring him the greater fame. He lost. Parkinson's Law, in the more general form of 'work always expands to fill the time available', soon flew around the globe, while Price's sank without trace. He later observed that for a while there was a Price's Law in metallurgical physics, but his friends knew that it was little consolation.

Still, Price was now hooked on the idea of studying how scientific knowledge had accumulated, and he couldn't wait to apply his physicist's mind to the history of science. He left Singapore and enrolled for a second PhD at the University of Cambridge. Reflecting his love of lab equipment, his thesis was to be on the history of scientific instruments. He felt that these measuring devices – from the microscope to the oscilloscope – were the key to scientific advance. Rutherford couldn't have split the atom without the accelerators he used to fire particles at each other, while Einstein had relied on recent experimental results to come up with the equation that described the energy locked up in those atoms: $E = mc^2$. Right back to the seventeenth century, supposedly the birth of modern science, Price felt that the real credit should not go to the gentlemen scientists playing with their new toys and discussing their latest observations over dinner or at the Royal Society. It belonged to the unsung instrument-makers, who blended technical ability with science and applied skills handed down over centuries to craft precision equipment for their wealthy customers. These were the men who determined not only what questions could be asked, but what could be discovered, and Price resolved to tell their story.

It was 1950, the year that Ellen gave birth to their first child, Linda. For a while Price thought he could measure that too, bringing graph paper to his wife's bedside and noting down the periodicity of her contractions, so that he might predict the birth-time of the baby. When the infant failed to arrive on time he felt let down and angry. Nature ought to work this way! What kind of god would choose messy randomness over cool, elegant predictability?

Settling in at Cambridge, Price found plenty of role models, not least Joseph Needham, who was the West's leading expert on the history of Chinese science. Needham also came from a scientific background – he was a biochemist at Cambridge until a young female Chinese student, Lu Gwei-djen, arrived in his lab in the mid-1930s, taught him the language and awakened in him a passion for China. (Half a century later, after his wife died, he finally married her.) Needham taught Price the necessity of knowing everything there was to know about one's chosen subject. Instead of being content with English sources, as most Western historians were, you had to read everything that had been written, whether it was in German or Chinese or Arabic. And if you didn't speak Chinese or Arabic, then you had to find someone who did, and work with them until the meaning of each passage became clear.

Price put his training to good use. While sifting through medieval documents in the old library of Peterhouse college looking for references to scientific instruments, he came across a manuscript that struck him as odd. Written scruffily on parchment in dark brown ink, *The Equatorie of the Planetis* contained instructions for the construction and use of a medieval astronomical device called an equatorium. Based on similar geometric principles to the astrolabe, the rarer and more complicated equatorium showed the positions of the five planets known at the time on its flat disc, as well as those of the Sun, Moon and stars. The manuscript had been in the library since 1542 and had long been attributed to an astronomer called Simon Bredon. But Price noticed references in the tables to astronomical observations made in 1392, and Bredon died in 1372. He couldn't have written it.

Astronomy texts of the time were generally written in Latin, and Price had only ever seen one other Middle English text on the subject. It was a lesson in the use of the astrolabe by one of the biggest names in medieval literature, Geoffrey Chaucer, the author of *The Canterbury Tales.* Although best known for his poetry, Chaucer was fascinated by the stars and often worked astrological references into his stories. He probably wrote *A Treatise on the Astrolabe* – his only known non-literary work – for his son, Lewis. A small footnote hidden in the tables of the equatorium manuscript tempted Price to a conclusion almost as bold as his exponential law. It read 'radix Chaucer'.

'Radix' refers to the reference date that an astronomer uses to compare all of his observations against, hinting that the author of the text and tables could have been none other than Chaucer himself. As he studied the text and analysed the language it used, Price became convinced. Chaucer had indeed written the manuscript as a companion piece to his treatise on the astrolabe, telling the story of the planets' motions as he narrated the first of his *Canterbury Tales*. What's more, the messy corrections suggested that this wasn't a scribe's copy – the manuscript was an original in Chaucer's own hand, the only extensive sample of his writings in existence. It was a sensational claim. But no one has since been able to refute Price's analysis, leaving experts still arguing today about whether he was right.

This extraordinary find diverted Price yet again, nudging him round one more twist in the path that was to lead to Athens, as he decided to specialise in the history of astronomical instruments. More than any other type of instrument, he was starting to realise that the earliest scientific devices related to the heavens – the equatorium, and before that the astrolabe, and the sundial. These were the devices that would lead him to the beginning of the instrument-makers' story. They told him that stretching back centuries, people had looked to the dancing lights in the sky and felt the same urge as he did – to measure, to understand, and to predict. He wanted to know where it came from, the knowledge encoded in these instruments, and to trace this will to understand that connects hundreds of generations of human beings who have wondered at the passage of the stars.

Around this time he collaborated with Needham and the Chinese historian Wang Ling on a survey of ancient Chinese astronomical clocks. Their findings appeared in the eminent British science journal *Nature* in 1956, squeezed between two reports from the frontier of scientific progress: an outline of plans for what turned out to be the first successful crossing of Antarctica, and a summary of a conference at which biochemists excitedly discussed the likely mechanism of DNA, the structure of which James Watson and Francis Crick had stunningly announced in the journal just three years earlier.

Price's paper, on the other hand, was from another time and another world. It described a Chinese tower from the eleventh century AD. According to a text written from the period by an imperial tutor called Su Sung, the tower housed a huge astronomical clock driven by running water or mercury. The liquid poured into scoops on a big wheel, which each took the same period of time to fill before they weighed enough to trigger a release mechanism and drag the wheel round another step. As well as telling the time, the ornate, ten-metre pagoda displayed the movements of the heavens.

It must have been an awesome sight. On the roof was an armillary sphere: a huge three-dimensional version of an astrolabe's sky map that would have weighed 20 tons, with a little Earth in the middle and metal rings around it representing the equator, horizon and equinoxes. Just beneath, on a covered platform, sat a great globe marked with the star constellations. The mechanics of the clock drove both around in time with the rotating sky, while within the body of the tower a series of brightly painted puppets were attached to the central revolving shaft, so that they appeared through doors at the front and sounded the time with bells and gongs.

Historians had got the development of clocks all wrong, argued Needham, Price and Ling. The conventional view had been that throughout most of history people told the time using non-mechanical devices such as slow-burning candles, hour glasses, sundials and simple water clocks. Then mechanical clocks appeared out of nowhere across Europe in the thirteenth century – cogs and all – when some genius invented the escapement. The escapement is the bit of a clock that turns the continuous energy from whatever is powering the mechanism (say a falling weight or an unwinding spring) into a series of discrete steps of equal time – the clock's 'ticks'. In one simple mechanical escapement, for example, a pendulum weight rocks back and forth in a way that allows a gearwheel driven by the clock to move by one tooth each time the pendulum swings.

The appearance of mechanical clocks in Europe is seen as one of the most crucial moments in the history of technology. The first known examples were extravagant astronomical affairs, which appeared in the thirteenth century and spread throughout the continent during the Renaissance. Like Su Sung's water clock they displayed the motions of the heavens, telling the time more as an afterthought. It was only later that these clocks were stripped down and miniaturised into something that might be more familiar to us today.

This development was so important because these clockmakers with their skills in precision gearing were crucial in developing the automated devices that ultimately made possible the machinery of the Industrial Revolution (as evidenced by the fact that we still call any sort of automated gearing 'clockwork'). Take the differential gear, for example. It's a sophisticated arrangement of gearwheels in which there are two independently driven inputs. These two wheels are connected by a pinion in such a way that their relative motion drives a third wheel at a speed that is related to the difference between the two inputs. It first appeared in Europe in the planetary displays of Renaissance clocks, but then the idea was adopted in textile mills to regulate the speed that cotton threads were wound onto a bobbin, depending on the tension of the strands that were feeding it.

The invention of the differential gear allowed cotton yarn to be mass-produced faster, cheaper and better than it could be by hand, and it revolutionised one of the most economically important industries of the period. Then the arrangement was reversed (allowing one input to drive two independent outputs) for use on a road steam engine, to make the powered wheels more steerable. Eventually it was adopted in automobiles, and is still used today.

Western historians had credited the development of all this to Europe. Exactly where the first clockwork clock came from, and why the earliest ones were so sophisticated (instead of progressing from simple to complicated, as the history of technology is supposed to do) was a mystery. But once clockwork clocks appeared, the foundations for the modern age had been laid.

It was thought that such timepieces then spread from Europe to the rest of the world: Jesuit missionaries took them to China, for example, in the sixteenth and seventeenth centuries. But Western historians had neglected to consult the appropriate Chinese texts. If they had, they would have found that much of the knowledge involved in the mechanical clock was already known in China. These texts contained descriptions of a series of increasingly elaborate astronomical clocks, culminating in Sung's. It wasn't mechanically driven, but the wheel nonetheless acted as a kind of escapement, slowing and controlling the continuous flow of the water into a series of regular time steps. And simple gearwork had driven the mannikins and spheres round at the appropriate speed. Historians would need to look much further back for the true origins of clockwork.

The Chaucer work on the equatorium and the *Nature* paper boosted Price's career and reputation, not to mention his ego. But during his research he realised that his biggest discovery was yet to come. He had learned about another astronomical object that was far older and far more complex than the equatorium or the Chinese clocktower or indeed than any other known instrument. Price read the papers of Svoronos, Rados and Rehm, and realised that no one had yet come close to unravelling the mystery of the Antikythera mechanism.

It wasn't clear exactly what the device was, but he could see that the gearwork it contained was more sophisticated than anything else known for more than 1,400 years, at least until the complex astronomical clocks of Medieval Europe. All of the various strands that Price had been tracing back through history – scientific instruments, astronomical knowledge, clockwork – were becoming entwined. And at the head of them all stood the Antikythera mechanism. Price believed that this unique object held the secret to the origins of the entire technical tradition that had led to the first clockwork clocks and then to the advances that ultimately enabled the scientific and industrial revolutions.

The mechanism turned upside down conventional ideas about the scientific legacy of the ancient Greeks. Historians had tended to write them off as clever up to a point, skilled in philosophy and sculpture, but not practically minded. Now, here was evidence that they had been masters in mathematical gearing; that they had built a clockwork computer more than a millennium before anyone else had even dreamed of such a thing. Price realised that the Antikythera mechanism was the oldest surviving trace of a technology that had been crucial to the emergence of the modern world.

Questions bounced about in his head. What had happened to the technology? Could he prove a direct link to modern clocks? How could the Greeks have developed such sophisticated technology without leaving any trace in the historical record? And since they did, what else might they have been capable of? But first, he had to find out more about the mechanism itself.

In 1953 he wrote to Christos Karouzos, the director of the National Archaeological Museum in Athens, to ask for more information. Karouzos duly sent him the most recent photographs of the fragments, which showed that since the publications of the 1920s and 30s, cleaning had revealed several previously hidden features. Price couldn't understand why such a revolutionary find was largely ignored by historians and archaeologists, and he wrote a couple of articles about it, including one in the British science magazine *Discovery*, in April 1957.

'If it is genuine, the Antikythera machine must entail a complete reestimation of ancient Greek technology,' he urged. 'Its discovery 55 years ago . . . was as spectacular as if the opening of Tutankhamen's tomb had revealed the decayed but recognisable parts of an internal combustion engine.' But it was impossible to glean much detail from the black and white images Karouzos had sent. With so many questions surrounding the mechanism, Price was unable to support his grand claims with anything more than colourful words.

And so, in the summer of 1958, he has come to Athens.

He used his considerable charm to persuade Karouzos to let him study the Antikythera fragments directly, and despite a slight bafflement as to why this eccentric British scholar was so keen to see these shabby fragments when the museum held items of much greater importance and beauty, the director merely shrugged and agreed. Now, in a basement storeroom – just a few metres away from the oblivious crowds admiring the Antikythera Youth so glamorously displayed in the museum's echoing exhibition hall – Price finally comes face to face with another side of ancient Greece, a forgotten strand of invention whose threads he sees woven into everything around him, in every car and bicycle, every clock and calculator.

And like Valerios Staïs before him, he is transfixed by its alien, crumbling appearance. Unlike Staïs, however, Price sees much more of the mechanism's workings where the external coating of limestone has been scraped away: the big four-spoked gearwheel, that wouldn't look out of place on a bicycle; the engraved dial, with a graded scale no different to those on the voltmeters that fascinated him as a student; and the smaller wheels behind, not unlike those in the watch on his wrist.

He turns the pieces over and over, looking for any tiny clue that the previous researchers might have missed. Day after scorching hot day Price returns to the museum, always making a beeline for the storeroom. He scrutinises every visible detail of his treasure, and measures it and counts it and notes it down. In particular he checks its ragged edges, trying to work out how the different fragments would have fitted together.

A Greek epigrapher called George Stamires also happened to be in Athens that summer. Price charmed him, too, into helping him to translate the legible parts of the inscriptions, especially the bits that were uncovered by the most recent round of cleaning. Whereas John Svoronos had read 220 letters, and John Theophanidis upped that to about 350, Stamires was able to decipher almost 800. The writing dated to the first century BC.

Price concluded from his measurements that rather than being disparate fragments from a larger mechanism, the pieces all fitted together, meaning that they comprised a greater part of the complete device than had been thought. And he realised that the various dials and plates in the flat fragments had not been squashed together and distorted by the weight of the seawater, as previous scholars had feared. The wheels within the fragments were actually very nearly in their original places – the mechanism had all along been quite flat. It should be possible to start looking at how the individual gearwheels had actually worked, rather than just making vague guesses at the machine's overall purpose.

Like John Theophanidis, Pericles Rediadis and Albert Rehm before him, Price believed that the mechanism was originally kept in a rectangular wooden case; it would have looked, he imagined, like a well-made eighteenth-century clock. On the front was a big central dial, almost as wide as the case itself, and on the back were two dials of the same width, one above the other. There were traces of small doors on the front and back, made of flat bronze plates. On every remaining surface of the doors and the front and back faces were inscriptions engraved into the bronze.

Price turned his attention to the front dial, which had two scales around its edge. Only the bottom portion of the dial had survived, but by counting the gradations, he could estimate how many divisions each scale had originally contained.

The inner scale was divided into 12 sections of 30, adding up to 360. In the very bottom segment of the dial Stamires was able to make out a complete word: *Chelai* (ΧΗΛΑΙ), meaning 'claws', and the ancient Greek name for the zodiacal constellation of Libra. The claws belonged to a giant scorpion – the body of which forms the next sign, Scorpio – waiting to swallow the Sun as it passes the autumn equinox and into the winter sky. One step counterclockwise on the dial, just two legible letters . . . NO . . . were enough to suggest the name of the preceding sign, Virgo, which the Greeks called Parthenos ('virgin'), after their goddess Athena. Here, then, was confirmation of what Rediadis had suspected when he saw the word μοιρογνωμονιον in the machine's inscriptions half a century earlier. The mechanism's scale had shown the 360 degrees of the zodiac, with the twelve signs running clockwise around its edge. A pointer edging around this dial would have traced the Sun's annual journey through the heavens.

The outer ring was divided into 365 segments, with the month name Pachon (ΠΑΧΩΝ), as spotted by Rehm, and the first two letters of Payni (ΠΑ . . .) visible in the surviving top portion. These were two consecutive months of the ancient Greco-Egyptian calendar, which was divided into 12 periods of 30 days each, followed by an extra period of 5 days to make up the 365 days of a year. This scale must have shown the months of the year, again running clockwise round the dial. While the pointer's position on the inner dial showed the Sun's path against the background stars, the outer scale would have given the date.

This particular calendar was the favourite of astronomers throughout the Hellenistic world, because every year had exactly the same sequence of months and days, with no adjustment for leap years. This guaranteed that everyone meant the same thing by the same date. But it had the disadvantage of being slightly shorter than the actual solar year, which is 365 and a quarter days long, so it shifted with respect to the seasons by one day every four years (as would our calendar if we didn't slot in leap years). Accordingly, the outer ring appeared to be rotatable. The user must have been able to move it around by one day every four years, to keep the calendar in sync with the zodiac scale.

Etched onto the scale were tiny solo letters, not evenly spaced but in alphabetical order around the dial. Their meaning became clear from studying the other inscriptions on the front face. Only snatches of a few of these lines were visible, and they said things like: 'Vega rises in the evening', 'The Hyades set in the morning' and 'Gemini begins to rise'.

These newly translated morsels were familiar enough.

Such text was common on a type of calendar used by the Greeks from the fifth century onwards called a parapegma. They were a bit like primitive weather forecasts; their purpose was to correlate repeating astronomical events such as the risings and settings of particular constellations with phenomena on Earth, like weather patterns or the flooding of the Nile. Such calendars became the main method by which people marked the passing of the seasons, as well as providing invaluable information for farming and navigation.

Parapegmata probably developed from simple inscriptions that listed astronomical events – such as the dawn rising of the star Sirius – alongside the expected weather for that time of year. Later, these got a bit more sophisticated and they would be engraved on a stone tablet, say, with a peg hole alongside each item on the list. That way you could move the peg forwards by one hole every day and always know what time of year it was, without having to observe the stars directly.

The inscriptions on the Antikythera mechanism clearly served a similar purpose. They didn't have peg holes, though. Instead, reference letters were inscribed at the appropriate points on the zodiac scale. When the Sun pointer reached a particular letter, the user could refer down to the appropriately lettered item on the list below.

The text even gave Price a clue to the mechanism's origin. The parapegma with the closest wording we know to the one on the Antikythera mechanism was written by the ancient astronomer Geminus, who made his observations on the island of Rhodes.

As Virginia Grace and her colleagues would later discover, the ship on which the Antikythera mechanism was found almost certainly stopped off at Rhodes shortly before she sank. Here, Price found his own evidence of a link with the island. It's not certain when Geminus lived, but most scholars plump for the first century BC. Geminus's astronomy wasn't all that impressive – his writings mostly summarised the work of others for future students – but he was likely to have been on Rhodes at the time that the Antikythera ship sailed.

Price kept looking. On the back of the mechanism were two dials, one above the other. Each seemed to consist of a series of concentric rings, perhaps five on the top and four on the bottom, which were divided into segments of about six degrees each. Strings of letters and numbers were inscribed within these segments, though it wasn't clear what they meant. Each of the back dials also had a miniature dial embedded in its face, off-centre like the second hand on an old-fashioned watch. The inscriptions on the back were even more fragmented than those on the front. But even from the few legible words translated by George Stamires, Price could get an idea of the subject matter. They said things like 'two pointers, whose ends carry', 'the Sun's rays', 'ecliptic', 'Venus' and 'protruding'. As previous scholars had suspected, these inscriptions seemed to form an instruction manual for the mechanism.

Although he wasn't sure what the back dials were for,

Price guessed from the inscriptions that they had something to do with demonstrating the cyclical relations of the Moon, the Sun and maybe even the planets. As the wheels turned the dials must have calculated their respective movements through the sky, just as astronomical clocks did many centuries later. The Antikythera mechanism might not have shown hours and minutes, but nonetheless, he argued, it was concerned with 'time, in its most fundamental sense, measured by the wheeling of celestial bodies through the heavens'.

Beyond such sweeping statements, however, Price had no idea what the back dials would have shown. And despite his early optimism it was even harder to work out what was happening inside the mechanism. At least 20 gearwheels survived in the fragments, all cut from a flat sheet of bronze about 2 mm thick. In the middle of the mechanism was a flat bronze plate, with trains of gearwheels leading above and below it. They were driven by an axle that came in through the side of the case and turned a little gear that was positioned parallel to the side of the box (at right angles to all of the other gearwheels). This 'crown' wheel engaged with the big four-spoked wheel that drove all the other gears.

But there the trail ran cold; the intricate workings of the machine remained frozen deep inside those stubborn, calcified lumps. Unless he could reconstruct the internal mechanism, Price couldn't prove any of his conclusions – they were merely speculation, based on a few barely legible words.

And not knowing how the gearwheels functioned was immensely frustrating. At last Price had the Antikythera mechanism in his hands and he had begun to understand its purpose, but the 'particular go of it' remained a mystery. He wrapped the pieces carefully inside an old cigar box, tucked it out of sight at the back of a shelf, and admitted defeat, at least for the time being. The knowledge encoded in humanity's oldest surviving machine was hidden from him.

After his intense summer in Athens, Price took a position at the prestigious Institute for Advanced Study at Princeton in New Jersey. Again he was surrounded by brilliant scientists, many of them Europeans who had fled the Nazis in the years before the Second World War. He just missed Albert Einstein, who worked at the institute until his death in 1955, but elsewhere on the leafy campus was the mathematician Kurt Gödel, who was grappling with places that Price's graphs could never reach: the theoretical limits of knowledge and beyond. The scientific historian Otto Neugebauer, though officially based at nearby Brown University, also spent a lot of time at Princeton and shared with Price much of his enormous knowledge of ancient astronomy.

Price didn't think much about Gödel – whose ideas about the limitations of mathematics didn't sit well with Price's rational, measurable world view – but he was inspired by the institute's director, Robert Oppenheimer. During the war, Oppenheimer had been the scientific leader of the Manhattan Project, which succeeded in developing the first nuclear bomb. As director, Oppenheimer was razor-sharp but impatient, jumping from topic to topic and staying with each just long enough to grasp key questions and confound the experts who had worked in the field all their lives, before he moved on to new ground. Critics tutted that he never concentrated on one subject long enough to make the progress that a physicist of his brilliance should have been able to achieve. But Price admired the audacity of it, and felt that he and Oppenheimer had a lot in common.

One of the first things Price did at Princeton was lecture on the Antikythera mechanism and his conviction that it held the key to the origin of modern machines. News of Price's work soon reached the author Arthur C. Clarke (another of Price's heroes, along with H. G. Wells), who had recently moved to Sri Lanka. As well as writing science fiction, Clarke was a keen diver and had just published several books about underwater exploration. It is unclear how he first heard about the Antikythera mechanism – the memory was lost to him in the later stages of his life; perhaps it was from Jacques Cousteau, with whom Clarke attended the first US skin divers' convention in Boston in February 1959. But when he did finally learn of this mysterious artefact, he felt that Price was on to something of fundamental importance.

Clarke introduced Price to Denis Flanagan, the passionate editor of *Scientific American*. Flanagan duly persuaded Price to write an article about the mechanism and it appeared as the magazine's cover feature in June 1959. Again, Price called for a complete rethink of the history of technology: 'Nothing like this instrument is preserved elsewhere,' he said. 'On the contrary, for all that we know of science and technology in the Hellenistic age we should have felt that such a device could not exist.'

Price sent a copy of this article to Clarke. 'Please find some more,' he wrote hopefully at the top. But neither Clarke nor anyone else has ever found anything comparable (Clarke once recalled that the most advanced artefact he had ever found as a diver was an early nineteenth-century soda-water bottle).

After two years at Princeton, Price took a job at Yale as the university's first professor of history of science. Price set about filling his department in New Haven, Connecticut, with graduate students, as well as scientific instruments. Antiques from past centuries, mostly wood and brass, decorated every room of the department and of his suburban home. He would tinker with them, as he had with the physics equipment he tended so faithfully as a student back in London, and he soon became known as the man to whom instruments spoke. Like Virginia Grace with her amphoras, Price could coax a story out of any mysterious mechanical object, gleaning how it worked and what it did from subtleties in the design that others barely noticed.

Yet he couldn't make any progress on the Antikythera fragments. Over and over he studied the drawings and photographs he had brought back with him, and he visited Athens again in 1962 to check his readings and confirm that the pieces fitted together as he thought. He even tracked down

Albert Rehm's unpublished notes on the device, which had been kept in Munich since Rehm's death after the war. But he still couldn't work out how the gears functioned, and the further cleaning that he had insisted on wasn't progressing as he had hoped. The archaeologists Virginia Grace and Gladys Weinberg contacted him after seeing his article in *Scientific American* to ask if he might publish a reconstruction of the mechanism alongside their work on the other contents of the Antikythera wreck. Humiliatingly, when their paper was published in 1965 he could still add nothing to his previous study.

Worse, his theories about the mechanism had made little impact. One account of his work, an article in the Athens press by a senior American professor, even scoffed that Price had been fooled by the layers of corrosion into thinking that the Antikythera mechanism was much older than it really was. It was a planetarium, said the professor scornfully, similar to one with which he had been taught the layout of the solar system as a child in school in Austria some 60 years before. It had clearly fallen on the site of the Antikythera wreck by chance, many centuries later.

Such ridicule stung sharply. Price often woke in the night, staring at the ceiling and wondering whether he could be wrong about the mechanism, whether for all the evidence he had been taken in by some cruel hoax. Was he wasting his reputation on a dead end? But during daylight hours Price allowed himself no such indulgence and kept busy on other projects. Mercurial, his colleagues called him, and they could only watch breathlessly as he rode from one topic to the next on a wave of a childlike enthusiasm. Like Oppenheimer, Price revelled in becoming an instant expert on everything and telling those who had spent their careers quietly mastering a subject exactly where they were going wrong. Whatever anyone else told him they studied, he felt a compulsion to join them – to see what they saw, do what they did, and to better it.

He was soon giving both historians and sociologists a run for their money. By counting the number of scientific papers published in different fields, and analysing precisely who was citing who, Price extended his theories about the growth of science. Traditional historians were still distinctly unimpressed by Price's claims, which they felt were simplistic and driven more by a love of the dramatic than any true understanding of social development or the accumulation of knowledge. They hated how he ignored inconvenient data points for the sake of a good story, although their criticisms also contained a hint of snobbery. Then Price stopped bothering with historians and started talking to scientists. They loved his work – at last, instead of advancing fuzzy opinions someone was studying science with numerical methods that made sense to them. His ideas about the growth of science have since been cited in journals from aeronautics to zoology. And in March 1965 Price was afforded one of science's highest honours, an invitation to give a lecture at the Royal Institution in London.

Price's theories may not have seized the popular imagination like Parkinson's Law, but he helped to lay the foundations of a whole new field of study: scientometrics, the science of science itself. He concluded that science had grown by five orders of magnitude (more than 16 doublings) in the three centuries since the foundation of the Royal Society, meaning that '80 per cent to 90 per cent of all the scientists who have ever lived are alive now'. He argued that the pattern of recent citations among the world's scientific papers could reveal the areas where research was actively progressing, not to mention the relative importance to science of particular journals, authors, institutions and even countries. And he declared the secret of distinguishing science from non-science: the higher the proportion of citations of newer papers (those less than five years old) compared to older ones (more than 20 years old), the more likely that an article is scientific.

Price himself believed, as usual, that he was uncovering universal truths about the nature of knowledge and where it was taking humanity. Little green aliens coming to Earth would understand the Planck constant, the velocity of light, or the wave equation no matter how much they differed from us. Surely, he mused, they would also recognise his scientometrics.

Hoping all the time that he might unearth further clues to the ancient mechanism that started it all, Price also followed up his interest in the history of astronomical instruments and clocks. With the help of his students he studied and catalogued all of the ancient sundials and astrolabes he could get his hands on. And in 1967 he persuaded *National Geographic* to pay for him to go to Athens to investigate the Tower of the Winds, in return for writing an article about it in the magazine when he got back.

The octagonal tower is one of the only buildings from ancient Greece or Rome that has never been buried or demolished, or even lost its roof. It was built at the beginning of the first century BC, around the time of the Antikythera mechanism, by a Macedonian astronomer called Andronicus Kyrrhestes. On its faces are carved eight winged demigods – one for each of the eight winds – their intense yet flowing features still as evocative as ever. Etched beneath them are eight sundials, with webs of lines that showed both the time of day and the season from the direction and length of the shadows cast. Missing today is the bronze weathervane in the form of Triton, son of the sea god Poseidon, that once swung with the wind to point to the appropriate divinity: Lips, for example, god of the south west wind, which used to carry ships into Piraeus, Athens's port.

The inside of the tower, by contrast, has been completely gutted. It was used as a church in early Christian times and then, during Turkish rule, as a prayer site by Muslim dervishes, whose whirling dances brought them closer to God. In the 1760s two British antiquarians dug down through the centuries of trodden dirt and bones and found the original floor. Carved into the marble was a mysterious pattern of holes and grooves. Some huge and complicated equipment had once stood there, they concluded, and guessed it was some sort of water clock. In Roman texts the tower was referred to as an *horologium*, which means 'hour indicator'. And the ancient name of the spring that runs above the tower in the hill of the Acropolis is Clepsydra, which literally means 'water thief' and was a name often used for water clocks.

No archaeologist had ever attempted to suggest how the water clock might have worked, since its mechanism had completely disappeared. But Price was confident he could solve the mystery. Deciphering what the floor markings had once supported, he reckoned, was like 'recreating the workings of a suburban kitchen in an empty room, using the relative positions of the sockets, pipe holes and rectangular floor stains as evidence'.

His extensive knowledge of ancient Greek water clocks was essential – much of it from writings of the Roman architect Vitruvius in the first century BC. These clocks didn't have the complex clockwork mechanism and escapement of modern clocks; instead, a regular flow of water measured time as it passed. Vitruvius described two basic types of clepsydra made by the Greek engineer Ctesibius. The simplest, common in Egypt since about the third millennium BC, consisted of a water vessel with a hole in the bottom, which measured time as the water level dropped. It wasn't very accurate, because the flow rate depends on the weight of the water above, so these clocks slowed down as the water level fell. Ctesibius invented a better version in the third century BC that was adopted throughout the Greek and Roman worlds. Water poured into a container that was engineered to keep a constant water level – either by an overflow pipe near the top or by a ballfloat that blocked the inflow pipe when the tank was full, a bit like the ballcock in a modern-day cistern.

Water then dripped from a hole in the bottom of the container at a constant rate into another cylindrical vessel. The rise in this water level over the day was used to measure the passing hours. With each new dawn, the tank was emptied and the clock was started again. Price felt sure that the Tower of the Winds was a giant version of just such a clock.

Price and his colleagues – a photographer and his wife, and a draftsman sent by *National Geographic* – spent days painstakingly clearing the floor of the tower; moving the marble rubble, sweeping away the dirt, and cleaning out the drain holes with a plastic-handled potato peeler, until they had an exact floor plan of the tower and a small cylindrical chamber that adjoined it.

In the smaller chamber they discovered that the stone joints of the floor had been reinforced by lead-coated bronze clamps. Something heavy must once have rested there. Nearby, a discoloured groove ran up the wall and there was a rectangular hole in the floor. Price concluded that this chamber must have housed the clock's main water tank. The groove in the wall probably held a lead pipe, carrying water under pressure from the nearby stream up into the top of the water tank. The hole in the floor must have been where the tank was emptied at the end of each day.

A pipe from the tank would have carried water at a constant rate to the measuring tank in the main part of the tower. Here, channels in the floor showed where overflow from the water tank fed three fountains. Price also found grooves that he calculated held railings to keep spectators away from the clock's machinery. He even located some battered panels that fitted them among the heaps of marble that lay nearby. But there were some grooves that Price couldn't interpret. He did what he usually did in such circumstances – decided that the failing was not his, but the ancient stonemason's, who had clearly carved the marks in the wrong place by mistake.

It wasn't possible to tell from the floor and wall markings how the time had been displayed. An hour pointer attached to the rising float or a hammer striking a gong at set times would have been in line with clocks of the period. But there was another, more spectacular alternative that Price felt would have been appropriate for the awesome Tower of the Winds.

At the turn of the century, two ancient pieces of bronze discs had been found – one in Grand, north-east France, and one by builders digging foundations for a house in Salzburg, Austria. Both were inscribed in Latin and were found with Roman remains, dated to around the second century AD. Albert Rehm studied the Salzburg fragment and published a reconstruction of it in 1903, just a few years before he first saw the Antikythera mechanism in Athens. He concluded that the complete disc must have measured more than 60 centimetres across, and that it had been the face of a large astronomical clock. The fragment was inscribed with figures representing the constellations – robed Andromeda, her naked husband Perseus with sword held high, Auriga the chariot driver, as well as the zodiac's Pisces, Aries, Taurus and Gemini.

Rehm realised that the disc he was studying matched another type of clock that Ctesibius had designed, which provided an elegant astronomical solution to the problem of the seasons. In ancient Greek and Roman times, hours weren't all the same length. Day (measured from sunrise to sunset) and night (sunset to sunrise) were each divided into twelve equal hours, the length of which therefore varied throughout the year. This complicated the hour markers on clocks somewhat. One way to get around this was to have different hour plates for different seasons, or even to carve curved hour lines on a cylinder that was turned a little each day.

But the Salzburg clock was basically a water-powered astrolabe. The bronze disc, marked with the constellations in the sky, was set vertically on a central axle behind a fixed set of curved wires that represented the position of the horizon and the hours of the day. The float in the water clock was connected to the axle so that as the float rose the disc turned. From the front an observer would see the stars and constellations riding clockwise through the sky, mirroring the movements of the heavens.

To account for the changing seasons, a series of holes had

been punched through the disc in a circle to represent the ecliptic, the Sun's annual path through the sky. Each hole represented the Sun's position on a particular day. When a peg depicting the Sun was placed in the appropriate hole, the time could be read off as it sailed past the static hour lines. The ecliptic circle was offset from the centre of the disc so that in summer the peg hole would take the Sun high through the sky during the day with only a relatively short path below the horizon, the opposite being the case in winter.

Price imagined such a clock as the centrepiece of the Tower of the Winds. The gleaming bronze star disc turning inexorably and mysteriously in line with the sky must have been the main attraction of the bustling Athens marketplace. This was much more than just a timepiece. It was a spectacular celebration of the beauty of the heavens, and of man's understanding of it. And he became convinced that although it contained no gearing, the clock was closely associated with the Antikythera fragments in scientific detail and in spirit; the notion of representing the skies with a flat disc that turned with the heavens must surely have served as an important inspiration for whoever thought up the two-dimensional dials and pointers of the Antikythera mechanism.

Another success for Derek de Solla Price! But his satisfaction didn't last long. Arthur C. Clarke still wanted him to publish a reconstruction of the Antikythera mechanism – especially since going to see the fragments for himself.

Clarke attended an astronautics congress in Athens in the summer of 1965, where American astronauts were celebrating their successful return from the *Gemini 5* mission. It was the first space flight to last eight days – the length of time it would take to get to the Moon and back, and hopes were high that they would reach the Moon before the decade was out, as President Kennedy had promised. Man was no longer just observing the heavens, he was conquering it.

Clarke took time out from the party to track down the old mechanism he had heard so much about. It took reluctant museum staff several days to locate the fragments in their cigar box, and he was dismayed that such an important relic was not on display. But when he finally got to unwrap the pieces of battered clockwork, it was worth the wait. He saw immediately that what Price had been telling him was true. This was surely the most important single item to come from ancient Greece, and one of the greatest mechanical inventions of all time.

Yet coming face to face with such familiar technology was also strangely disturbing. It was the clearest possible demonstration that the Greeks were like us, with minds like ours. Where Price saw the continuing threads, Clarke also realised how much had been lost. It was unsettling to think that in the Antikythera mechanism the Greeks had come so close to our modern technology, only to fall back again for so long. He articulated his thoughts a few years later in a lecture on the limits of technology at the Smithsonian Institution in Washington DC. If the Greeks had been able to build on their knowledge, Clarke told his audience, the Industrial Revolution might have begun more than a millennium ago. 'By this time we would not merely be pottering around on the Moon. We would have reached the nearer stars.'

5

A Heroic Reconstruction

The Moon, which is the last of the stars, and the one the most connected with the Earth, the remedy provided by nature for darkness, excels all the others in her admirable qualities.

— PLINY THE ELDER

NO MATTER WHAT story you try to tell about the twentieth century, in the end you find its course diverted by the Second World War – a great, dark smear on history that sucks in everyone and everything before releasing them, a few years later, on new trajectories. Even Price's graphs show a blip. The supposedly inexorable growth of knowledge hangs suspended, just for a moment, before the curve begins its steady climb once more.

Some stories are only affected a little, some are pulled far off course. But every one is changed. While the Antikythera fragments are buried under Athens, so Albert Rehm endures forced retirement in Munich, Virginia Grace misses her amphoras while exiled in Cyprus, and Derek de Solla Price teaches a gruelling schedule of physics in London. All their futures depend on physicists in the United States and

Germany, who are locked in a race to unleash the devastating power of the atom. The outcome will determine their subsequent paths, which will in turn lead to unforeseen collisions and crossings, opportunities that fall into place one by one, a trail of influences spreading outwards like a chain reaction and stretching into a future that will soon feel to those living it as if it could never have been otherwise.

Their trajectories are set at 5:29 a.m. on 16 July 1945, when the efforts of the 130,000 Americans working on the Manhattan Project finally come to fruition in the middle of the New Mexico desert. Twenty miles away, the physicist Richard Feynman ignores the official advice to use dark glasses, figuring that the truck windshield will protect his eyes from the radiation, and thus he becomes perhaps the only person to see the full force of the explosion. He watches the fireball turn silently from blinding white to yellow to orange before black smoke curls around its edges and grows into a cloud so dark it looks as though a hole is being ripped out of the sky. A minute and a half later, the silence is broken by a tremendous bang that steals his breath and shakes him to the bone. The Atomic Age has begun.

Price played his part in our understanding of this era. For decades the accepted story had been that only Germany and the United States had been trying to develop nuclear weapons. US officers arriving after the war to see the dishevelled remains of Japan's premier physics laboratory saw no reason to believe that it might once have been the site of a Japanese Manhattan Project, and the scientists they interviewed said nothing to challenge that view.

But with the help of his Japanese graduate student Eri Yagi Shizume (and still following Joseph Needham's advice about going beyond what's written in English), Price uncovered unpublished historical records and diary entries which showed that Japan had indeed been pursuing its own nuclear bomb in what it called Project Aeropower. Yoshio Nishina, the country's top physicist, had been friends with Albert Einstein and Niels Bohr in Europe during the 1930s and when war broke out, the Japanese Government told him to make a nuclear bomb. He had been in the middle of building a pilot plant to concentrate the uranium-235 needed to sustain a nuclear chain reaction when the facility was bombed in an air raid in April 1945.

In 1971 Price collided once more with the fallout from that nuclear arms race, and this time it handed him a key to the Antikythera mechanism. With Arthur C. Clarke's encouragement he had held firm to his conviction that understanding the infuriating device was of fundamental importance for everything that he had been studying. His reconstruction of it would be his biggest achievement; rewriting the history of technology, if not of our entire civilisation. But there just wasn't enough information on the surface of the delicate fragments to see how the gears worked. Then came a breakthrough. Price saw a technical report that had been published a few months earlier by researchers at Oak Ridge National Laboratory in Tennessee. It described how gamma rays from radioactive isotopes could be used to peer inside metallic objects of artistic or archaeological importance without destroying them. His long wait was over. Now it was excitement, rather than frustration, that kept Price awake at night.

He wrote to Alvin Weinberg, the director of Oak Ridge, to ask if he could use the new imaging technique on the Antikythera fragments. Oak Ridge was one of three labs originally set up as part of the Manhattan Project, and Weinberg had played a senior role in that effort. While Robert Oppenheimer oversaw the design of the bomb at Los Alamos, New Mexico, it was Weinberg's job in Tennessee to purify uranium-235 and work out how to produce plutonium from uranium (a process that was then carried out on a larger scale at the third site, near Richland in Washington). Back then Oak Ridge had employed a staggering 40,000 people, but it was now home to just a few thousand physicists, whose job it was to build on the knowledge gained during the war for peaceful ends, from medical imaging to nuclear power. Weinberg became a vociferous champion of the latter – so much so that when the reactor at Three Mile Island suffered a partial meltdown in 1979, he argued that it actually proved how safe the technology was, since the situation had ultimately been brought under control.

The United States wasn't the only country keen to harness the power of the atom after the war. Having witnessed its world-changing potential, pretty much every government that could afford it set up an agency with similar aims, Greece included. So when Weinberg received Price's letter he put him in touch with the Greek Atomic Energy Commission. The trail led to nuclear physicist Charalambos Karakalos, the head of radiography at a nuclear research lab in Athens. Price turned on his charm once more and explained what he wanted, but Karakalos was sceptical about the chances of success. His lab was still under development and was only equipped with the most elementary tools for radiography. And no one had ever tried to image anything as corroded as the Antikythera fragments – it wasn't at all clear that there was even any structure left inside to see.

Still, the project sounded more interesting than anything else he was working on at the time, so he headed across town to the National Archaeological Museum with a small sample of radioactive thulium-170 and some radiographic film. The stable form of the element, thulium-169, has 69 protons and 100 neutrons in the nucleus of each of its atoms. But the unstable thulium-170 has one extra neutron squeezed into each nucleus. One by one, its atoms undergo radioactive decay, spitting out electrons and high-energy photons (also known as gamma rays) in the process. Thulium becomes ytterbium and erbium. The number of thulium atoms left halves every 128 days, as regular as clockwork – the exact reverse pattern to the one Price had once seen traced out in books against his bedroom wall.

Karakalos set up a rudimentary darkroom and took a series of exposures of the largest piece of the mechanism.

He knew that the photons emitted by the thulium would shoot through the fragment and strike the film behind it, breaking up silver bromide crystals trapped in its emulsion into ions. Any metallic atoms inside the fragment would block the photons, leaving an invisible shadow of intact molecules on the film.

Breathing softly in the dim glow of the safelight, Karakalos took the transparent, green-tinted film and delicately placed it into a bath of developing solution; this would convert the exposed silver ions into black, metallic silver atoms. And there it was. A picture that made 2,000 years pass in an instant. As the film turned black he saw jagged green shapes left behind; the outlines of precisely cut gearwheels that tumbled into view one on top of the other, rendering the sophisticated handiwork of their long-dead creator visible at last. Not that the level-headed Karakalos put it that way. 'The images are of fair quality,' he noted. 'They show some new gears in fragment A.'

Karakalos went back to his lab and returned with two portable X-ray machines and a lot more film. X-rays are photons, too, kicked out of atoms when electrons are fired at an element such as tungsten. The X-rays from these machines were of lower energy than the gamma-ray source, meaning that he could use longer exposure times, and so control the amount of radiation hitting the film more precisely. Over the summer of 1972 he used them to take hundreds of images of the mechanism, painstakingly adjusting the focal distance, angle and exposure time – anything up to 20 minutes for each – to get the sharpest possible pictures of what lay inside those ragged, irregular fragments.

Price was on sabbatical in Europe that summer and he visited Athens twice to look over the radiographs with Karakalos, checking progress and studying the features of the mechanism that the images revealed. The crucial details Price needed were how the wheels were arranged – which ones meshed with which – and the number of teeth on each. This would allow him to work out the numerical ratios that were encoded by the gear trains and therefore to establish once and for all what the mechanism had been designed to calculate.

Karakalos's wife, Emily, helped with the tooth counts – Karakalos thought she would produce the most accurate numbers as she had no preconceptions as to what numbers to expect. Every day, after laying an X-ray image on the light box in front of her, she would run her palms over its surface as if to smooth away imaginary dust, and angle a magnifying glass precisely over each wheel in turn. Closing her mind to noise and other distractions, she attended only to the tiny green zigzags, counting up the visible teeth on each and noting down the results. For the smaller wheels she used enlarged black and white prints created from the negatives, drawing a careful circle to mark the circumference of each gear and perforating each tooth tip with a pin from her sewing box. Then she'd turn the print over, to number the holes in neatest pencil on the back.

It was tedious work. All of the wheels in the mechanism

appeared on top of each other in the images, up to eight layers deep, so many of the details were blotted out. Karakalos did his best to vary the exposure times and focal distance to isolate the details of each one, but even so there wasn't a single gear for which every tooth was visible. Determining the total number of teeth, therefore, meant counting those that could be seen, measuring what angle of arc was visible, then scaling up to the entire 360° circumference. It was easy to make mistakes – the teeth on some of the wheels were actually quite irregular and often it wasn't clear where the centre of a wheel was, leaving its exact size in doubt. The counts had to be repeated again and again, on print after print, until a consistent figure was established for each wheel.

Sometimes Emily became distracted, wondering at the foreign professor for whom these jumbled patterns meant so much. His enthusiasm could be infectious, but she had never seen anyone whose mood changed so quickly from one day to the next. It was especially hard to predict how he would react when he saw the results of her work. Some days he was approving and encouraging; others he would frown and demand a recount. She couldn't understand why he could not be satisfied with the evidence from the images her husband had so carefully produced and that she had so painstakingly counted. Why look at all, if you cannot accept what you see?

When the counting was done, Price went back to Yale, shut the door to his office, and continued to work feverishly on a reconstruction of the mechanism. As well as confusing the tooth counts, the fact that all the wheels were superimposed in the images made it difficult to tell which gears meshed with which – it was hard to distinguish a gear at the front of the mechanism from one at the back, for example. He made a model to help visualise the mechanism's workings; its two cardboard faces held on to four wooden sides with a buckled, cotton strap. He drew on the outlines of the existing fragments, then added his reconstruction of the front and back dials, complete with little cardboard pointers. Inside, he arranged and rearranged the cardboard wheels like miniature furniture.

First, the easy part. Price confirmed that the small crown wheel drove the mechanism by engaging with the big four-spoked wheel, which he grandly called the 'Main Drive Wheel', because it then drove the rest of the machine's gear trains. The shaft of the crown wheel extended out through the side of the case. Price was undecided how the crown wheel itself would have been turned – by hand via a handle on the side of the case, he concluded, or even by a spectacular water clock like the Tower of the Winds.

The drive wheel was positioned directly behind the zodiac dial on the front of the case and turned around the same centre. It takes the Sun one year to go all the way through the zodiac, so Price deduced that this big wheel would have driven a pointer that showed the position of the Sun in the sky. Five turns of the side handle would have moved the wheel and pointer through roughly one turn – a time period of one year.

From there, things got a bit more complicated. The rotation passed through three connected pairs of meshing gears, ending up at a gearwheel that turned around the same centre as the main drive wheel, but with a narrower axle that passed back up to the front of the mechanism, through the middle of the drive wheel's hollow shaft. This presumably drove a second pointer around the front dial.

What did the second pointer show? To say for sure, Price needed to know the speed at which it moved relative to the Sun pointer. And by counting the teeth on the gears he could calculate what happened to the speed of rotation at each step. For example, as described in Chapter 2, if a wheel with 20 teeth drives a wheel with 10 teeth, then every turn of the first wheel will result in two turns of the second. This can be written mathematically.

$$\frac{20}{10} = 2$$

And in a similar pair of meshing gear wheels, say with 90 teeth and 30 teeth, every turn of the first wheel will result in three turns of the second.

$$\frac{90}{30} = 3$$

The two pairs can be connected together by a common axle, running through the second wheel of the first pair and the first wheel of the second pair. Because they share an axle, these wheels have to turn at the same rate, meaning that the output of the first pair becomes the input of the second pair. This can be written as follows:

$$\frac{20}{10} \times \frac{90}{30} = 6$$

In other words, for every full turn of the first wheel, the final wheel of the four will turn six times. Of course it would be easy to achieve such a simple result with a single pair of gearwheels – with 60 teeth and 10 teeth, say – but combining two, three or more pairs of gears allows for more complicated ratios than can be easily achieved with a single pair (smaller wheels are simpler to cut than big ones, for example, and it's harder to get the spacing of the teeth right with prime numbers).

Price sifted through the tooth counts that Emily and Charalambos Karakalos had provided for the six wheels in the first gear train, trying to work out what the overall ratio had been. What was the ancient maker trying to compute? The obvious answer, of course, was that the second pointer had shown the motion of the Moon. But he needed to prove it.

Calculating the position of the Moon from that of the Sun isn't straightforward. Although the calendar we use today divides years into exactly twelve months, the Moon doesn't go round the Earth exactly twelve times for every time that the Earth goes round the Sun. So a simple calendar can either show the timing of the Sun and seasons, or the motions of the Moon. But it can't do both – the two soon get out of sync. Our modern system of counting the days is driven by the Sun, and our calendar matches the seasons in a yearly cycle. This means that on a particular date each year the Sun will be in roughly the same position relative to the Earth. January is always winter (in the northern hemisphere at least) and July is always summer. The summer solstice – the longest day of the year, when the northern hemisphere is tipped maximally towards the Sun – falls unfailingly on 20 or 21 June.

The price we pay for following the Sun is that our calendar has lost touch with the Moon. The date of the full moon varies from month to month, and every year the pattern is different (this is why Easter, which is calculated according to the first full moon after 21 March, moves around in the calendar). The days of our months conveniently follow the same pattern every year – we know that March will always have 31 days, and April will always have 30 – but they no longer reflect the lunar phases.

Nowadays it doesn't matter that much. For most of us the Moon's phase is of little importance in daily life. But for the Greeks, as for most ancient peoples, following the Moon's motion was critical for everything from the timing of frequent religious festivals and civic duties to whether you'd be able to see that night.

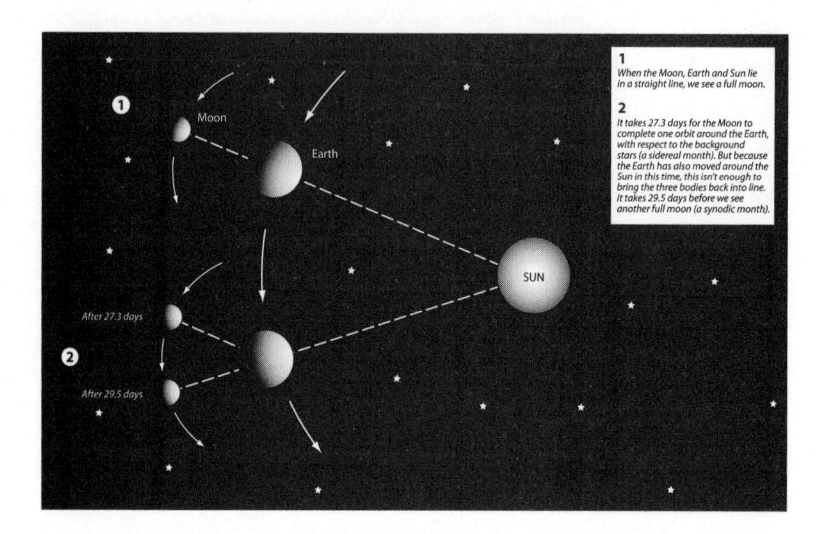

The Moon orbits the Earth – from our point of view circling through the sky with respect to the background stars – about every 27.3 days. This is called the sidereal month (from *sidus*, which is Latin for 'star'). The time period from full moon to full moon, called the synodic month, is a little longer, on average 29.5 days. The Greeks knew that although the Moon's movement doesn't fit neatly into a year, it does come back to almost exactly the same position with respect to the Sun and the Earth every 19 years. Within each 19-year cycle, there are 235 synodic months (give or take a couple of hours) and the Moon circles through the sky 254 times.

So the Greeks combined the movements of the Sun and Moon in a 19-year repeating calendar called the Metonic cycle, after an astronomer called Meton who lived in Athens in the fifth century BC. He was the first Greek we know of to use it, although he almost certainly got the idea from the Babylonians. Their priest-astronomers had been observing the heavens for centuries before that, and were well acquainted with the relationship.

According to this cycle, the number of sidereal months in one year is 254/19. So Price realised that once you have a wheel that turns with the Sun through the sky, you can multiply its rotation by this ratio to calculate the speed of the Moon. The counts that Emily and Charalambos had given him for the six wheels in this train were 65 (although they couldn't rule out 64 or 66), 38, 48, 24, 128 and 32. That gives the following gear train:

$$\frac{65}{38} \times \frac{48}{24} \times \frac{128}{32} = \frac{260}{19}$$

It is tantalisingly close to the 19-year cycle. Price played around with the numbers, hoping that the ratio he needed would appear. Changing the first wheel to 64 teeth – at the lower end of the range that the Karakaloses had given – brought the top number to 256. Then all it took was to tweak the 128-tooth wheel to 127, surely within the range of possible error. The train now read:

$$\frac{64}{38} \times \frac{48}{24} \times \frac{127}{32} = \frac{254}{19}$$

Price sat back in his chair and lit his pipe, sucking on the end as he watched the particles of smoke dance in the light of his desk lamp. The mechanism was giving up its secrets to him at last! And they were beautiful. The results of centuries of astronomical observations had been converted first into mathematics and then made real again, carved quite precisely into six wheels of shining bronze. The gear train reminded him of a computer programme: put in the Sun, and get out the Moon. As the owner turned the handle on the side of the box – driving the main wheel and the Sun pointer round once for every year – the second pointer on the front dial would have shown the Moon's position in the sky, whirling around the zodiac scale just over twelve times faster than the stately Sun.

But there was a snag. Each time one gearwheel meshes with another, the direction of rotation is reversed. So the gear train Price had just worked out, with its three pairs, would have sent the Moon spinning in the opposite direction to the Sun. That couldn't be right. But Price soon came up with an ingenious solution. Rather than the main drive wheel carrying the Sun pointer, he calculated that there must have once been a second wheel of the same size just in front of it, now lost, driven by the other side of the crown wheel. This would have turned at the same speed as the main drive wheel, but in the opposite direction, so this second wheel must have carried the Sun pointer, around the same way as the Moon.

This wasn't the end of the gearing. It seemed to Price that the two speeds of rotation achieved so far – corresponding to the movements of the Sun and the Moon in the sky – each fed further back into the machine, into a cluster of gear-wheels that were mounted on a bigger turntable. Price was stumped . . . until he had a crazy idea.

When he reconstructed the missing clock in the Tower of the Winds, Price had succeeded where others failed by looking at things from the point of view of the ancient craftsman he was trying to second-guess. The stars above the lit streets of Connecticut aren't as bright as they would have been in ancient Greece, but now he looked up anyway, watching the ghostly silver crescent of the Moon wax and then wane as the stars completed their graceful arcs behind it. Each fresh moon was like a new life, its cycle by far the most dramatic thing in the night sky. The maker of the device would surely have wanted to capture it.

Calculating the phase of the Moon is basically the same as working out the number of synodic months that have passed. If you start at full Moon, for example, then after each whole number of synodic months that passes, the Moon will be full again. For every half number of synodic months, the Moon will be new, and so on. The number of synodic months in any time period is intimately connected with the number of sidereal months and years, because the phase of the Moon depends not just on its position with respect to Earth, but the position of the Sun.

Imagine the Earth as the tip of the hour hand on a giant clock face in space, with the Sun at the centre. The Earth inches its way around the clock face as the Moon circles around the Earth in turn. At full Moon, all three fall roughly in line with the Earth in the middle, the Sun's rays shining past us to illuminate the Moon head on as we look at it. If the Moon then completes exactly one rotation around the Earth it will come back into the same position with respect to the background stars – say, from one o'clock to one o'clock. But because the Earth is itself sweeping around the Sun, one orbit of the Moon isn't enough to bring the three bodies back into line. The Earth is now at two o'clock with respect to the Sun. So the next full moon doesn't occur until the Moon has travelled the extra twelfth of a circle. In one year, these extra twelfths add up to one extra sidereal month. The relationship holds in general – the number of sidereal months in a particular time period is equal to the number of synodic months that have passed, plus the number of years. In one 19-year period, for example, 235 + 19 = 254.

The Greeks didn't necessarily think of it in such heliocentric terms, but thanks to the Babylonians and their 19-year cycle, they knew the numerical relationships involved. And just as you can add the number of years and synodic months that pass to get the number of sidereal months, so you can subtract years from sidereal months to get synodic months (for example, 254 - 19 = 235).

Price was looking at a gear train in which two speeds of rotation – representing the speed of the Moon, and in the reverse direction the speed of the Sun – were fed into a cluster of linked gearwheels mounted on a turntable, such that their relative motion had driven the turntable around. Two inputs, one output.

And so the answer came to him. It must be a differential gear – a set-up already familiar to him from the astronomical clocks of Renaissance Europe. Whereas the parallel gear arrangements he had noted so far could multiply and divide rates of rotation according to the ratios of their numbers of teeth, a differential gear could add and subtract.

Differential gears are complicated, a whole new level of gearwork, and to find one in such an old device made the Antikythera mechanism more astounding than ever. If Price was right there was no question of viewing the mechanism as an early stumbling attempt at mathematical gearing, the beginning of a technological line that might have died out as quickly as it began. To come up with a differential gear takes a virtuoso talent, both in mathematics and in practical execution, and it must have been the culmination of generations of experience.

There was already a hint that the differential gear was known in ancient times: a legend that around 2600 BC, the Chinese Yellow Emperor Huang Di had a chariot topped with a wooden figure that always pointed south. A differential gear could in theory achieve this by subtracting the revolutions of one wheel from the other, thus keeping track of any changes in direction. But it's probably just a story. No surviving text describes a working model until the third century AD, and there's no description of how it might have worked until the eleventh century.

The first differential gear known in the West – and the first used anywhere for a mathematical purpose – was in the eighteenth century. Its origins aren't clear, but it was possibly invented by the British watchmaker Joseph Williamson, who wrote in 1720 that he had constructed one for use in a clock, so that as well as showing the time, the mechanism could calculate the varying speed of the Sun through the sky.

The differential gear is an impressive invention because using one to make a calculation involves working out that the way in which its various parts move relative to each other is governed by a precise mathematical relationship. Two wheels driven independently of each other are both connected to a third wheel in such a way that it moves round with a speed that is half the sum of the speeds of the two input wheels.

In the Antikythera fragments Price thought he saw the remains of a triangle of three little wheels, all mounted on a bigger turntable. He had worked out that one of these wheels turned with the speed of the Sun, driven directly from the main shaft, and one of them turned in the opposite direction, with the speed of the Moon. The third wheel in the triangle was a pinion that connected the other two in such a way that as they turned relative to each other they pushed around the turntable on which they were both sitting. In effect, the motion of the Sun through the sky was being subtracted from its lunar equivalent. Multiply the resulting movement of the turntable by two and the machine was calculating the phase of the Moon.

Price followed the gear train on down and concluded that this rate of rotation was then transmitted to the rings of the lower back dial to show the 235 synodic months of the 19-year cycle, with the position of the pointer within each segment corresponding to the Moon's changing phase. The subsidiary dial would have showed the twelve synodic months of the lunar year.

That just left the upper back dial. He could see that it was a series of concentric rings with a subsidiary dial divided into four, but only part of the gear train leading to it survived. After playing around with the numbers he had, Price guessed that it had shown the months of a four-year cycle, presumably so that the user could keep track of the 365-day calendar as it shifted against the seasons. He didn't know what all the different rings were for. It didn't really matter. He had decoded the stunning differential gear, and at last he understood the essence of the Antikythera mechanism. It was, he announced, a 'calendar computer'. It calculated the movements of the Sun and Moon as seen from Earth, in order to track the days and months of the year and, through the parapegma text, to predict the corresponding positions of the stars.

Price had also uncovered some hints as to where the information encoded within the mechanism came from. It was intriguing, for example, that the 19-year cycle it used had come originally from the Babylonians. They used centuries of observations to come up with equations that could predict the positions of celestial bodies, which they saw as messages that communicated the activities and intentions of the gods. They compiled astronomical tables on clay tablets that contained endless progressions of numbers, each line recording the incremental changes in the position of the Moon, say, and using simple algorithms to extrapolate those movements into the future – like lines of computer code carved in stone.

For all their precision, the Babylonians showed little interest in how the solar system was actually arranged; to them the night sky was basically a light show. The Greeks, on the other hand, were obsessed with coming up with geometrical models of the heavens. They wanted to explain the celestial movements – what orbited what, and how – not simply to predict them. In fact, they weren't too bothered about detailed observations at all. The arrangement of the heavens was a philosophical matter and proposed models were judged more on beauty than precise correspondence to reality.

The accurate dials and pointers of the Antikythera mechanism, although certainly Greek, showed a reliance on numerical relations that seemed much closer to the arithmetical spirit of the Babylonians. Whoever had invented the device seemed to have combined the two traditions. They surely had links to the East.

Price wrote up his findings into a 70-page opus called *Gears from the Greeks*, which he published in June 1974. The Antikythera fragments already represented the oldest geared mechanism – or mechanism of any kind – in existence, and by far the most sophisticated device that has survived from antiquity. But the discovery of a differential gear was breathtaking. It combined astronomical knowledge, abstract mathematical understanding and mechanical skill in a way that would not be matched again until the Renaissance. Yet the Antikythera mechanism was made with an assurance and skill that made it all look easy.

Price remained convinced that this technology had not died out. When Greco-Roman civilisation collapsed in the early centuries AD, much learning, including that of mathematics and astronomy, was transferred first to the Islamic world, and in later centuries back to Europe. Price argued in his paper that a knowledge of the gearing of the Antikythera mechanism had been carried through to safety each time.

Among the evidence Price cited was a manuscript written around 1000 AD by an eminent Islamic astronomer called Abu Rayhan al-Biruni. It described a geared calendar that al-Biruni called the Box for the Moon, which could be fixed to the back of an astrolabe. In it a train of eight gearwheels calculated the positions of the Sun and the Moon in the zodiac, as well as the phase of the Moon. An astrolabe with a very similar calendar, made in Iran in the early thirteenth century, survives to this day (in the Museum of the History of Science in Oxford). Price argued that these Islamic instruments were direct descendants of the Antikythera mechanism, and that when the knowledge passed back to Europe it triggered the sudden flowering of astronomical clocks. This explained why the first mechanical clocks spread so quickly and involved such complex displays of the heavens – the technology for these displays had already existed for centuries.

Gears from the Greeks showed that by relying on the few technical manuscripts and artefacts that have survived into modern times, historians were far behind in their understanding of what the ancients could do. Technology that had been attributed to Europeans in medieval and Renaissance times and beyond had in fact been mastered by ancient civilisations. Price believed he had shown that it was this knowledge that had finally triggered the explosion of technological advance in Europe that led to our own modern civilisation. His decades of conviction had paid off. He was finally ready to rewrite history.

But things didn't quite happen that way. Certainly, specialists in the history of technology welcomed Price's paper. The German scholar Aage Drachmann, for example, complimented Price on his 'comprehensive and unimpeachable investigation', while his British counterpart John North concluded that 'the reader . . . can hardly deny that the mechanism is the most important scientific artefact from classical Greek times'. Arthur C. Clarke, too, continued to champion the importance of the Antikythera mechanism, and included it in his first series of *Arthur C. Clarke's Mysterious World*, filmed in 1980.

But beyond this handful of enthusiasts, nothing much changed. Ancient history was discussed and understood and taught just as it had been before. The Greeks were still regarded as philosophers, masters of ideas, but uninterested in technical expertise, while the more philistine Romans excelled at shows of strength such as amphitheatres and aqueducts, but didn't have the intellectual imagination to equal the thinking of the Greeks. Any credit for our own technical expertise was always to be given much closer to home, to the fathers of modern-day science in Renaissance Europe.

One problem was that the gap between the Antikythera mechanism and the next known mention of a geared mechanism was well over a thousand years. The device appeared to prove that the Greeks had invented clockwork, but to say that the surviving Islamic instruments were part of the same technological line required a huge leap of faith, whatever one's opinion of Price. And his paper, although brilliant when discussing the wider context and significance of the device, was tortuously hard to follow when it discussed the gearwork, with jumps of logic that were never fully explained. So although nobody challenged his conclusions, nobody followed them up either.

It didn't help that Erich von Däniken, the controversial Swiss author, had featured the Antikythera mechanism in *Chariots of the Gods?* (1968), in which he argued that alien space travellers visited the Earth thousands of years ago, giving ancient civilisations advanced technology such as batteries and uncorrodible metal, and inspiring much of their religion. The Antikythera mechanism, said von Däniken, was proof that the Greeks had technology they couldn't have developed on their own (in a later book he even developed the idea that aliens used the device in their spaceships to navigate by the stars).

Chariots of the Gods sold millions, becoming a worldwide phenomenon. And it put the Antikythera mechanism firmly into the category of oddball mysteries, rather than a finding to be taken seriously by mainstream historians. Even after Price had published his paper, the mechanism was seen as an inconvenient aberration, mentioned as an afterthought, if at all. The ultimate snub came when Price's mentor from his years at Princeton, Otto Neugebauer, published his huge, massively comprehensive *History of Ancient Mathematical Astronomy* in 1975. He relegated the Antikythera mechanism to a rather derogatory footnote. In the decades to follow, although the details of Price's reconstruction were generally accepted, the wider implications were pretty much ignored. Both judgments were to prove wrong.

Back at the Athens museum, the fragments were at least put on public display. The attitude of staff there didn't change much, but in 1980 the mechanism did catch the eye of Richard Feynman, who since watching the New Mexico bomb test had become one of the most famous physicists in America. He went to Athens for a few days' break from lecturing, and wrote to his family on 29 June from the side of the pool of the Royal Olympic Hotel. The day before he had visited the archaeological museum and seen so many art objects and statues that he got all mixed up and his feet started to hurt. He felt he had seen it all before, except for one thing 'so entirely different and strange that it is nearly impossible'. It was some kind of ancient machine with gear trains, he said, like the inside of a modern wind-up alarm clock.

When Feynman asked for more information, he was greeted with blank looks. 'In fact the lady from the museum staff remarked when told the Prof. from America wanted to know more about item 15087, "Of all the things in the museum why does he pick out that particular item, what is so special about it?"'

The Greeks must think all Americans to be terribly dull, he mused, after tracking down Price's work and finding out he was from Yale, 'only interested in machinery when there are all those statues and portrayals of lovely myths and stories of gods and goddesses to look at'.

But there was one man on whom Price's paper made a lasting impression. At the Science Museum in London's elegant South Kensington a 26-year-old assistant curator called Michael Wright read the 70 pages avidly from beginning to end. He was responsible for looking after the museum's collection of Industrial Revolution machines, and Price's words lit the spark of what was to become a lasting obsession. He, too, was fascinated by how machines were put together, and when he saw Price's detective work it was like a glimpse into an exciting new world. He wished the project had been his own.

A few things didn't quite make sense to him, however, such as why the maker of the Antikythera mechanism would have used something as complicated as a differential gear to calculate the phases of the Moon, when a simple gear train would have done the trick just as well. And he thought it was odd that inscriptions suggesting that the mechanism might have shown the movements of the planets, which Price had discussed in his earlier *Scientific American* article, were now hardly mentioned. But Price seemed to have the whole thing sewn up. Wright put his confusion down to inexperience, filed away the paper, and got on with his work.

Gears from the Greeks was Price's last word on the Antikythera mechanism; he felt he had said all there was to be said on the matter. From that point on he looked ahead, to what he believed would be the next technological driver of knowledge: modern computers. Although most electronic computers at the time were slow grey boxes with the simplest of circuitry and just a few kilobytes of memory, Price predicted that the world was entering a 'computer age', in which the next step would be three-dimensional chips that allowed machines to jump to conclusions and think creatively – like people. Just as the linear arithmetical thinking of the Babylonians had given way to the three-dimensional geometry of the Greeks, computers would enjoy a similar evolution.

Only this time, instead of modelling the heavens, they would be modelling the brain. Computers more intelligent than humans weren't a huge step away, Price believed, and he saw such a scenario as wholly positive. Any resistance to the idea of super-intelligent computers was as backward as the Catholic Church's silencing of Galileo in the seventeenth century. Galileo used telescope observations as evidence that the Earth went round the Sun – contradicting the church's view that the Earth was at the centre of the solar system, which followed that of the ancient Greek philosopher Aristotle. But in Price's view, that wasn't why the astronomer was seen as such a threat. 'What was at issue was the validity of using a little bit of tubing with two bits of glass in it to attain knowledge that made you wiser than Aristotle and all the Church Fathers,' he said in an interview in 1982. 'Galileo was claiming that with an artificial device he knew things about the universe that the greatest minds of the past couldn't have known.'

Yet today we have no problem with the idea that a device can see what we can't; we happily rely on radio telescopes, X-ray imagers or particle accelerators and we believe in quarks, pulsars and DNA – all things quite undetectable by our naked senses. Just as Galileo's telescope improved on the eye, ushering in a whole new world of discovery, Price looked forward to the day when computers would improve on the brain.

He couldn't wait to dive into this new field of artificial intelligence, despite his health troubles after a heart attack in 1977. His family and friends urged him to slow down, but he couldn't imagine a life without work, or travel. In September 1983, having just undergone surgery after a third heart attack, he flew to London to stay with his old friend Anthony Michaelis – the editor who had published his first piece on the Antikythera mechanism in *Discovery* more than 25 years before.

One evening during his stay, the two were due to have dinner with Michaelis's girlfriend, Stefanie Maison. She planned to meet Michaelis during the day to get food for that night's meal, but in the morning he called and in an odd voice told her not to buy too much. 'Derek's not coming,' he said.

During the night, Price's heart had given up for the last time. As the stars in their constellations shone high above London's skyline, down below the light of a truly original mind was extinguished.

6

The Moon in a Box

The gods confound the man who first found out
How to distinguish hours! Confound them too,
Who in this place set up a sundial,
To cut and hack my days so wretchedly
Into small portions!

— MACCIUS PLAUTUS

JUDITH FIELD BURST triumphantly into Michael Wright's cluttered office. It was lunchtime at the Science Museum in London and, as usual, Wright was eating sandwiches at his desk and catching up on his reading.

Wright looked after the museum's engineering collection and Field was his opposite number, responsible for its astronomical instruments. She often used to come and sit in his office, drinking his tea (he made very good tea) and trying to outsmart him. Although he was meant to be responsible for contraptions from the time of the Industrial Revolution, Wright shared her interest in astronomical devices – the older the better – and they used to test each other's knowledge with newly discovered objects or snippets of information.

But today was different, Wright could see that. Today she really had something special. She reached into a padded envelope that she had been carrying, pulled out four battered pieces of metal and placed them on the desk in front of him with a flourish. 'What do you think of that?!'

The pieces were dark and worn, but otherwise in good condition. The largest was a flat, round disc, about twelve centimetres across, with a hole in the middle. It had Greek inscriptions on one face – including what looked like a list of city names and numbers, some graduated scales and a second hole offset from the centre, surrounded by a circle of seven intricately carved heads. The next piece was a disembodied metal arm that seemed to fit the central hole, with a hoop at one end for hanging the instrument vertically, rather like an astrolabe. And then the surprising bit. Two little axles, between them carrying four toothed wheels and a ratchet. The largest wheel, about four centimetres across, had two penny-sized circles cut out of it, and more writing around its edge.

As Wright turned the pieces over in his hands, Field related how she had come by them. A Lebanese man had walked in off Exhibition Road – straight off the street – and approached one of the uniformed security guards in the foyer. He said in broken English that he had something that might be of interest to the museum and pulled the metal bits out of his pocket as proof. Seeing the ancient lettering inscribed on the pieces, the guard figured he had better take the foreigner seriously, and called Dr Field.

Even though the instrument was in pieces, Wright knew

Captain Dimitrios Kontos and his crew of sponge divers in their tiny boat at the wreck site off Antikythera in 1901.

A Greek sponge diver from the early 20th century, wearing traditional diving dress. The helmet was handmade from copper and brass and weighed some 40 pounds. The suit was made from canvas twill and India rubber and weighed 18 pounds. The diver wore two 20-pound lead weights, one on his front and one on his back.

Some of the treasures recovered from the Antikythera wreck in 1900-01, now held in the National Archaeological Museum in Athens.
(*Above*) Bronze portrait head from a statue of a philosopher, from the 3rd century BC.
(*Below left*) Bronze statue nicknamed the Antikythera Youth, from the 4th century BC.
(*Below right*) Marble statue of a crouching boy, half corroded where it was exposed to sea water.

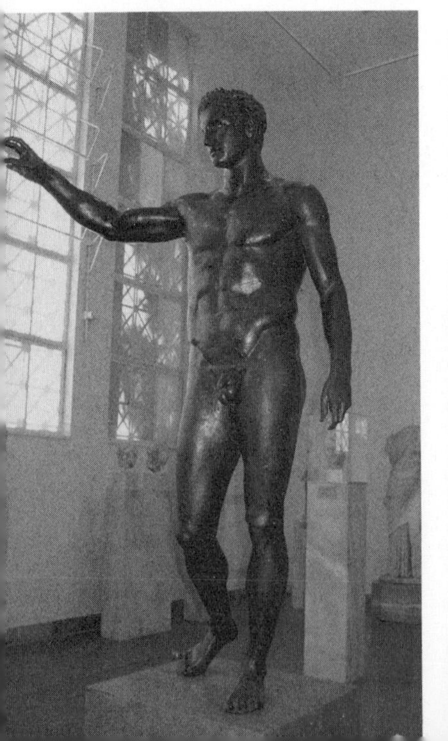

The three biggest surviving fragments of the Antikythera mechanism, on display in the National Archaeological Museum in Athens.
(*Above left*) Fragment C, showing part of the zodiac scale on the front dial.
(*Above right*) Fragment B, showing part of the upper spiral dial on the mechanism's back face.
(*Below*) Fragment A, showing the large 4-spoked gearwheel.

Derek de Solla Price with a model of the Antikythera mechanism in 1982.

The Tower of the Winds, a clocktower from the 1st or 2nd century BC, standing in the Roman marketplace in Athens. It once held a sophisticated water clock.

Allan Bromley with a reconstruction of the Antikythera mechanism at the University of Sydney. © Steven Siewert/ Fairfaxphotos

(*Below right*) Michael Wright with his reconstruction of the Antikythera mechanism, in his home workshop in Hammersmith, London, in 2006.

(*Right*) The parts of the model that drive the pointers for the planets. The epicycle discs and slotted levers for Mercury and Venus lie above the large wheel in the main assembly. The separate assemblies (from left to right) are for Mars, Jupiter and Saturn.

(*Below*) The main frame plate of Wright's model, with the large 4-spoked wheel and other gears attached to it.

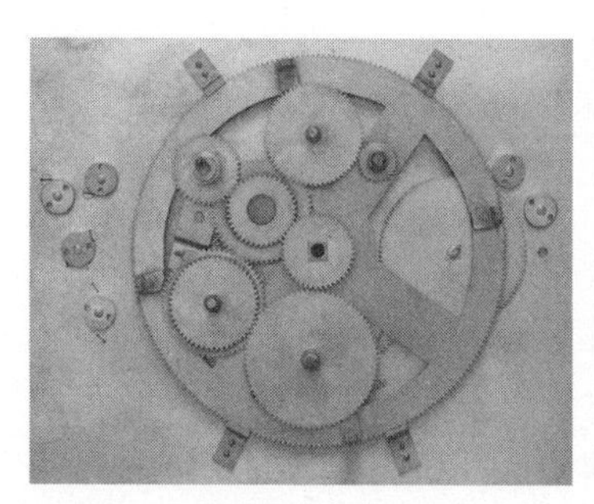

(*Above*) The Islamic geared astrolabe held in the Museum of the History of Science in Oxford (MHS inv. 48213). It was made in Isfahan, Iran, in 1221/2 AD. This is the back face, with a moon-phase dial (top), date indicator (top right) and concentric dials showing the Sun and Moon in the zodiac (bottom).

(*Below*) Pages from a 10th-century manuscript attributed to a Baghdad astronomer called Nastulus. The diagrams show how to make a similar geared calendar.

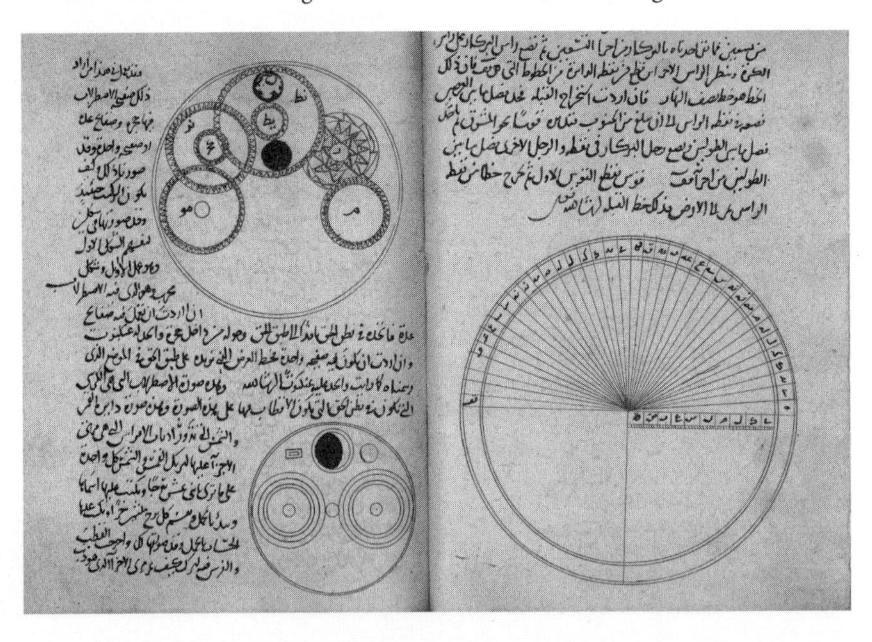

X-Tek's trip to X-ray the Antikythera fragments in Athens in October 2005. X-Tek's Alan Crawley connecting up the BladeRunner CT system, ready to X-ray the fragments.

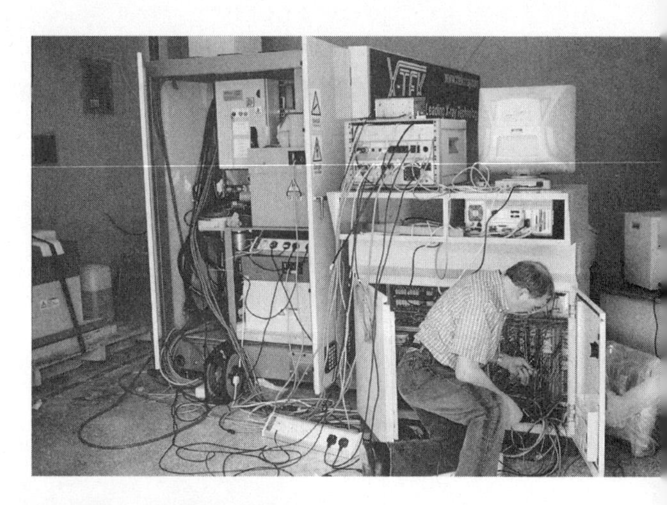

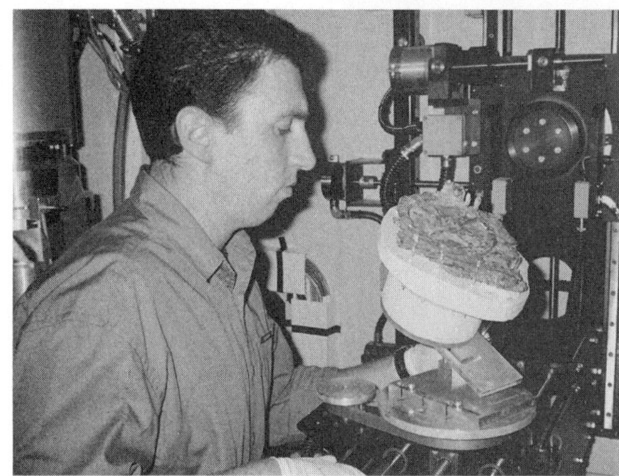

Pandelis Feleris, conservator at the National Archaeological Museum in Athens, loading fragment A onto BladeRunner's X-ray turntable for imaging.

Dr Eleni Magkou (National Archaeological Museum) and Andrew Ramsey (X-Tek) studying the first CT images of fragment G.

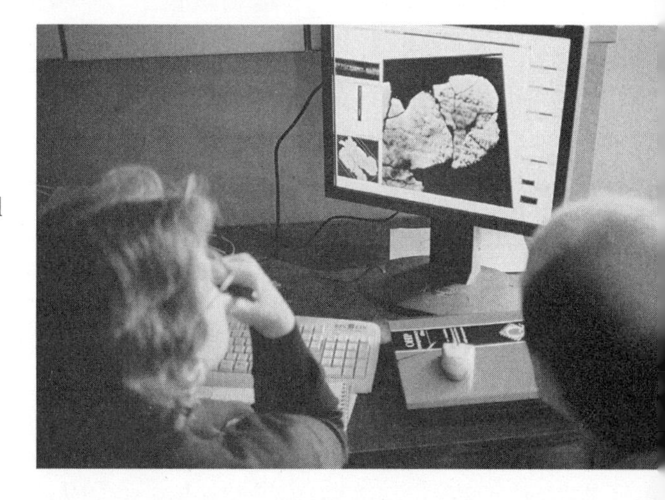

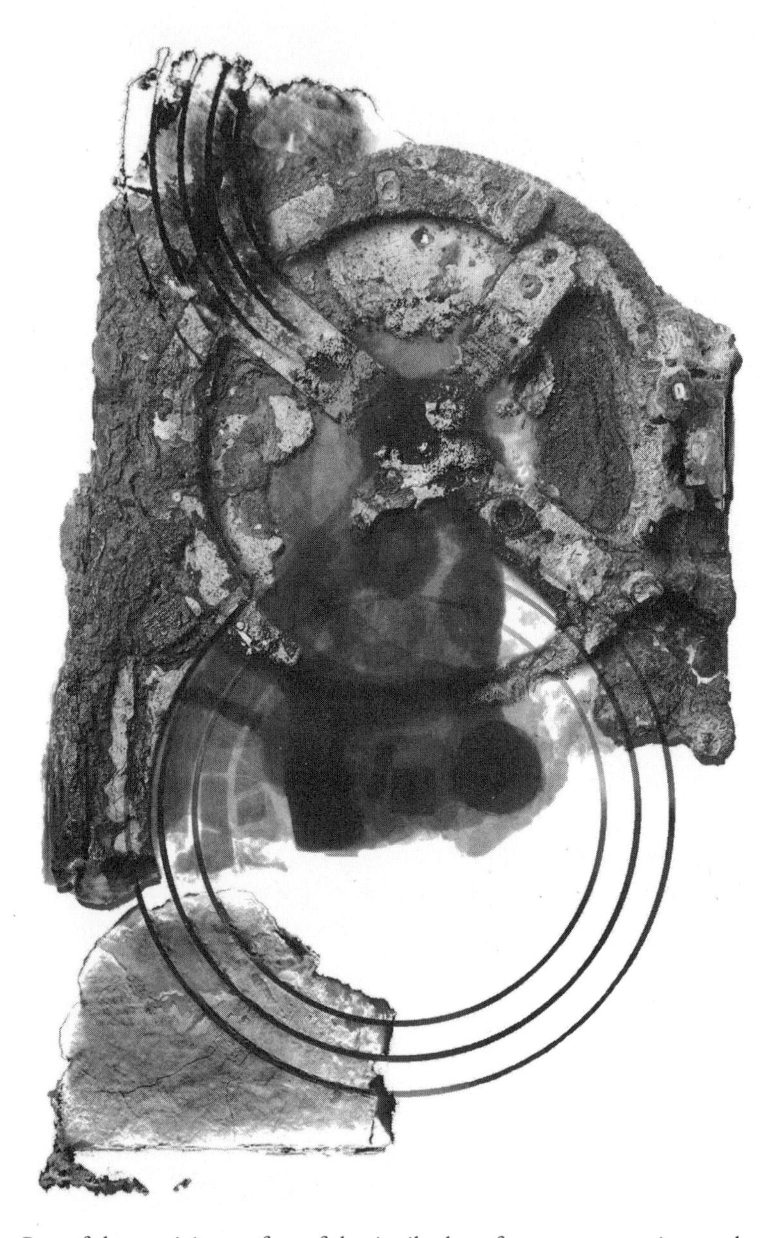

Part of the surviving surface of the Antikythera fragments, superimposed onto an X-ray image of the gearwheels inside.

that it was different from anything he had seen before. It was centuries old: the Greek inscriptions and the city name 'Constantinople' suggested straight away that it came from the Byzantine Empire – the Greek-speaking eastern territory created with Constantinople as its capital when the Roman Empire split into two in the third century AD. From the inscriptions on the main plate and the hanging arm he recognised it as a sundial. The remains of several similar instruments had been found from Hellenistic and Roman times. They were simple to use, although making them accurately involved some quite sophisticated knowledge of the movements of the Sun.

This type of sundial consisted of a disc with two scales round the edge, one for the latitude and one for the time of year. A bar called a gnomon fixed to the front of the flat disc by a pin in the centre, so you could rotate it, setting the angle depending on the latitude you were at. Then the arm slid around the edge of the disc to the appropriate time of year and the whole thing could be hung vertically, so that the bar was in line with the Sun. A stalk sticking out of the gnomon at one end cast a shadow, and you'd read the time off the hour lines engraved down the rest of the bar. The gnomon of this sundial was missing, but the latitude scale, month scale and list of cities and their latitudes, plus the hanging arm, were all similar to other ancient sundials that had been found.

But the gearwheels didn't fit at all. Sundials measure shadows; they have no need for wheels. And apart from the strange and unique Antikythera mechanism, no clockwork was known until the Islamic world. Between these two extremes there was nothing; a huge gaping hole for more than a thousand years. By the time the technology appeared again, the Greek and Roman civilisations had fallen, the Islamic world had risen, and Arabic was the new language of science. But this instrument wasn't Arabic. It was Greek.

Field had already made a rough count of the neat triangles.

'Look,' she said to Wright. 'This one has 60 teeth!' Field was a careful scholar, fiercely proud of her PhD; Wright didn't often get to tell her she was wrong.

'No,' he said, after a glance at the inscriptions. 'It has 59.'

The frown he knew was coming. 'How can you possibly know that?'

'It's numbered in Greek around the edge.' He traced his finger clockwise around the circumference of the wheel. 'Here, see, the numbers run 1 to 30, and then 1 to 29.'

Soon afterwards, Wright's supervisor, clock expert Rodney Law, came in for tea and the three chattered excitedly. By the time the pot was empty, one thing was clear. The 59-tooth wheel must relate to the movement of the Moon. On average, a synodic month (the time from one new Moon to the next) is about 29.5 days. So most ancient lunar calendars had months that alternated in length between 30 days and 29 days. If this gearwheel had turned by one tooth a day, then the numbers around the edge would have shown the day of the month.

The Lebanese collector had temporarily left the pieces with Field so that she could investigate the device. His intent soon became clear, however. He wanted to sell it the museum.

This presented a problem. The museum trustees were reluctant to buy anything that didn't have 'provenance', that is, an established paper trail showing where it had come from and who had owned it. There was no guarantee that items without provenance had not been stolen at some point, and the museum couldn't be seen to support the black market in smuggled artefacts.

But all the Lebanese owner could or would say about the pieces was that he had bought them a few weeks earlier from a street trader in Beirut (and even this story he later retracted). It was June 1983 and Lebanon was in the midst of a bloody civil war. Much of Beirut had been destroyed and thousands of its citizens killed under a hail of shells from the Syrian and Israeli armies. The city was in chaos and the idea that every artefact leaving the country should be accompanied by a tidy file of paperwork was laughable.

In the end the museum trustees decided that the geared sundial was too important to pass up; they couldn't risk it disappearing into a private collection before academics had a chance to study it. After checking with Interpol that no one had reported a similar device stolen, they handed over a hefty but as yet undisclosed sum, and the two curators had their sundial.

Judith Field worked on the origin and dating of the instrument. The list of place names and latitudes provided her main clues. Constantinople was at the head of the list, followed by towns and provinces including Alexandria, Antioch, Rhodes, Athens, Sicily and Rome.

Constantinople was founded in 324 AD when Emperor Constantine I renamed the city of Byzantium after himself, to serve as a second Rome at the gateway between the Mediterranean and the Black Sea. The resulting empire encompassed what are now Greece, Turkey, Syria, Lebanon and Israel, and northern Egypt. Its borders expanded and contracted over the centuries as it faced invasions from all sides – the Huns of Attila to the west, the Persians to the east, the Vandals to the south – and when the western half of the empire collapsed in the fifth century. By looking at the city names listed on the sundial and comparing the times at which they existed in the Byzantine Empire, Field worked out that the instrument must date to the late fifth or early sixth century, perhaps around 520 AD.

It became clear early on that the instrument was unique. As she searched the literature she realised that the sundial was the second oldest clockwork instrument known, preceded only by the Antikythera mechanism. There was no trace of anything similar after it until the geared calendar described by the Islamic scientist al-Biruni in the eleventh century. It was a heady feeling, to work on something that was so utterly isolated in the historical record.

Meanwhile, Michael Wright used his skills as a mechanic to deduce how the gearwheels must have worked, and to build a reconstruction. He had been fascinated by mechanical devices since he was a boy, when his parents would bring him to the Science Museum and let him loose with no instruction except that when the museum closed he should go down the road to find his uncle, who was an assistant keeper at the Natural History Museum. Back then the Science Museum was filled with old instruments in cases, like an enchanted forest of glass and metal. He'd invariably get lost, but always learned something new finding his way out.

At home and at school he had loved taking things apart to see how they worked, even if they didn't always go back together again. Soon he was building models: aeroplanes, trains, even working clocks. Like Price, he had studied physics at university, but his job now was to look after the exhibits in the museum's echoing energy hall – especially the hulking coal-burning steam engines that had powered Britain's Industrial Revolution.

He also took an interest in the museum's old clocks, including the stunning astronomical pieces that decorated Europe from the fourteenth century onwards. He was particularly fond of the museum's replica of one of the most impressive examples, completed by Giovanni de' Dondi in Padua, Italy, in 1364. One hundred and seven gearwheels drove seven golden dials showing the position in the zodiac of the Sun, Moon and five known planets: Saturn, Jupiter, Mars, Venus and Mercury. Wright came to know all the little tricks inside these clocks, and like Price he saw how the complexities of the gearing they used to model the planets turned up later in the steam-powered mills and looms of the Industrial Revolution.

Now, reconstructing the Byzantine sundial in his home workshop, he applied his skills to an artefact that no one had seen (or built) in many centuries. Some parts of the puzzle were missing, but he felt his way around the pieces, noting a scratch mark there, some asymmetrical wear here, trying to reconstruct how the device must have been put together. He quickly realised that despite the 500-year gap, the four surviving gearwheels had been part of a simple calendar very like the eight-wheeled one described by al-Biruni.

The gear ratios and layout of the Byzantine instrument were slightly different – and attached to a sundial rather than an astrolabe – but otherwise the principle was the same. Wright's reconstruction had a knob sticking up through the centre of the seven-headed dial, around which gods representing the Sun, Moon and five known planets stood guard over the seven days of the week. The ratchet, with seven lobed teeth, fitted directly underneath – clicking on exactly one day at a time as the knob was turned. As it rotated, a little wheel with seven teeth on the same axle drove the larger 59-tooth wheel, so that it turned one tooth for every day, and so that the numbers around its edge showed up one at a time in a little window on the missing back face of the disc – the day of the month. The two penny-sized holes in the 59-tooth wheel would have been filled with a black material, perhaps wax or dark wood. As the wheel turned, they moved under a circular hole on the back face to display the changing phases of the Moon. The missing wheels drove dials showing the positions of the Sun and Moon in the zodiac. The calendar was a nice little addition to the sundial, and clicking the ratchet on each day would have provided the date for setting the gnomon.

Once he had worked out the mechanism, Wright perfected two models of it. He filed the gear teeth until the wheels turned smoothly, then lovingly carved the Moon goddess's curls and topped both dials with fixing pins in the shape of horses' heads. The results were satisfyingly heavy in the palm of his hand as he ran his little solar systems through their paces, the latest in a long line of instrument-makers throughout history who have tried to catch the Moon in a box. Then he sent one of his models to his sons Gabriel and Caleb, who proudly showed it off to their friends at boarding school as the winter sun slanted over the playing fields (the horsehead pin, dropped and never found, is probably still there in the mud).

Wright had never been sure that he was doing the right thing with his life – he had ended up at the Science Museum more by accident than design. But now he felt a new purpose. Few of his colleagues as curators and historians could actually do what they studied and wrote about. He would combine his knowledge of ancient mechanisms with his practical skill as a craftsman to bring something unique to the field. By understanding the instruments from the maker's point of view he could glean insights that would be lost to a purely academic investigator. Maybe he was only a self-taught museum curator, but as a practical man he could still uncover things that the university professors might never see.

He also started thinking more and more about the

Antikythera mechanism. Price had suggested that there was a continuing tradition of geared instruments in the Hellenistic world and that this had been transmitted to Islamic culture. However, he had found no direct evidence to support either of those claims. Knowledge of astronomy and other subjects was certainly transmitted from the Greeks to the Arabs, but directly linking the Antikythera mechanism with a geared astrolabe from such a different world was a huge leap of intuition. Yet here out of nowhere was a stepping stone across a thousand years of history – a single shining drop of evidence that Price was right.

It now became clear that the Islamic clockwork calendars had not been invented from scratch. Their similarity to the one in Wright's sundial showed that they must have been influenced by the Byzantine tradition. And although the sundial calendar was much simpler than the Antikythera mechanism, its Greek inscriptions suggested that it was itself descended from Hellenistic instruments. What's more the workmanship – plain brass, without any traces of gold or silver, and gear teeth that were accurate but not excessively neat – suggested that this was an everyday object, not a rare luxury. There could have been hundreds, if not thousands, of these things all across the Byzantine world. In some simple form, at least, the tradition begun by the Antikythera mechanism of using gearwheels to model the movement of the heavens had survived.

But the more Wright studied Price's work, the more the details of it worried him. He was now ten years older and wiser than when he first read *Gears from the Greeks*, and he realised that many of Price's arguments just didn't make sense. In several places Price had cited the results of Karakalos' tooth counts only to reject them. For a gearwheel labelled E5, for example, Karakalos had estimated 50–52 teeth, but Price settled on 48 as more 'appropriate'. And for wheel G2, Karakalos counted 54 or 55 teeth, but Price had dismissed this as 'too small for any simple or meaningful interpretation of the gear train' and suggested 60 instead. Again and again, Price had changed the numbers for no obvious reason except that he needed to make them fit the gear trains already in his mind. He seemed to pull his ideas out of nowhere, and he made liberal use of supposedly practical arguments that to Wright, who had so much experience in making clockwork mechanisms, made no sense at all.

There was his old question of why the maker of the Antikythera mechanism would have used a complicated differential gear to work out the phase of the Moon when it could be done much more easily with a simple fixed gear train – as in the Byzantine sundial, for example. Wright also found Price's suggested function for the back dials ridiculously simple compared to the obvious sophistication of the device. Price had thought that the upper back dial showed a four-year cycle. But why would anyone go to all that trouble – a train involving seven gearwheels and a dial of five concentric rings – just to show a pointer that went round four times for every turn of the main wheel?

And then there was the big four-spoked wheel, the most

striking feature of the whole mechanism. Why was it so big and sturdy compared to the other wheels in the device? By grandly calling it the 'Main Drive Wheel' Price had given the impression that it was big and strong because it drove all of the mechanism's gear trains. But all that the wheel actually did in Price's reconstruction was to transfer its motion to a much smaller wheel turning around the same shaft, which then drove all the other trains from there.

Price's paper was certainly a stunning piece of detective work, but Wright felt he saw some sleight of hand in it as well. Even Price's name for the device – a 'calendar computer' – seemed designed to distract attention from the fact that his reconstruction of the mechanism corresponded to no known instrument, and didn't have any obvious practical use. For all of Price's insights, it was clear that he had barely begun to understand what the device was capable of. Wright wished he could discuss his questions with Price. The two men had briefly met, when Price had called at the Science Museum to see the newly discovered Byzantine sundial during his visit to London in 1983. But that was before Wright started studying the Antikythera mechanism in earnest, and within a fortnight of that meeting Price was dead.

Wright knew what he had to do. He would go to Athens to study the mechanism for himself and pick up where Price had left off. He would study the fragments, read the inscriptions, X-ray them for himself if necessary, and find out what the device really was.

But by this time the atmosphere at the Science Museum

had begun to change. It was under new management, which argued that the museum needed to focus less on studying obscure artefacts and more on pleasing its visitors or 'customers' as they became known. And Wright had a new supervisor who didn't approve of a curator spending valuable museum time on his own research projects. Wright's job was to look after the museum's exhibits – to maintain them, ensure they were displayed to best advantage and answer any public enquiries about them. Where, his boss asked him, did the job description say anything about flying off to Athens on a nice holiday to check out some obscure relic in the back room of a Greek museum? Wright's request for research time was turned down flat.

Kept busy on other projects, he dreamed instead. He was forbidden from spending any time on the Antikythera mechanism during work hours, but in his spare time he read up on ancient astronomy and technology, and brushed up on the Greek he had learned at school. When he finally got to Athens to see the mechanism for himself, he would be ready.

And then a bearded ball of energy rolled into Wright's office. His name was Allan G. Bromley.

Probing the nature of interstellar dust clouds was how Bromley had started his career, as an astrophysicist at the University of Sydney in Australia. That research got him interested in high-performance computing, and so he had become a lecturer in the computer science department. But he also had a keen interest in mechanical calculators, in fact anything to do with the history of computation and measurement.

In his small house in the Sydney suburb of Dulwich Hill, he had a garage full of adding machines, bits of clocks and giant analogue computers, one of them so heavy that it crushed the bricks on his driveway when he brought it home.

Bromley had first entered the Science Museum in 1979 on a year's sabbatical to study the notebooks and drawings of Charles Babbage, popularly known as the 'grandfather of computing'. The museum held the biggest collection of Babbage's papers, mostly abstract representations of his designs, such as logic diagrams and flow charts, and notebooks ('scribbling books', as Babbage called them) totalling thousands of pages. The scrawled notes were jumbled and fiendishly complicated, but Bromley was a worthy match for them. He knew the theory of computing inside out and had an excellent memory and insatiable desire for detail. He soon made himself a world authority on Babbage and became the first man to decode the diagrams since the author himself.

Babbage was famous for his attempts to build a machine that would construct whole sets of mathematical tables automatically. He, too, was inspired by a desire to predict the movements of the heavens. One day in 1821, at his house in Devonshire Street, London, he was proofreading hand-calculated astronomical tables with his friend, the astronomer John Herschel. He became frustrated by the numerous errors they found, and supposedly burst out: 'I wish to God these calculations had been executed by steam!'

'It is quite possible,' Herschel calmly replied. This got the 29-year-old mathematician thinking and within days he had come up with the idea for his Difference Engine. It took advantage of the fact that the path of any astronomical body can be calculated relatively simply by breaking it down into small chunks and adding the difference necessary to get from one step to the next – much like the arithmetic progressions used centuries earlier by the Babylonian astronomers. The resulting design was a huge contraption, involving hundreds of bronze gears, levers and wheels. Each digit of a number had its own wheel and the value of the digit was represented by the amount by which the wheel rotated. Babbage came up with a series of designs, culminating in the even more sophisticated and flexible Analytical Engine, which could multiply and divide, as well as add and subtract, store numbers, and even be programmed using punchcards. If it had ever been finished, it would have been the world's first programmable computer.

The British government paid Babbage £17,000 – a fortune at the time – to build his machines. Sailors relied on astronomical tables for navigation, so having a more accurate way to produce them would have been invaluable for a nation that relied for so much of its wealth on overseas trade, not to mention saving countless lives. But quarrels with his engineer and an inability to stop tinkering with his designs meant that Babbage never completed a single working machine. He died a bitter man.

After Bromley's sabbatical was over he visited London whenever he could to continue his studies of Babbage's papers, mostly in the British winter when his students were on summer vacation. His short stature, along with his bushy beard, rosy cheeks and yellow waistcoat (knitted by his mother), became a familiar sight at the museum, as well as in London's flea markets, which he regularly scoured for antique mechanical calculators and measuring devices that he shipped back home to his collection in Sydney.

By the mid-1980s Bromley was formulating a plan: the Science Museum would build one of Babbage's machines for the bicentenary of his birth in 1991. Bromley was convinced that the inventor's designs would work, but he had been looking at them from the point of view of a computer scientist. He understood the logic and the theory behind them, but he wanted to know more about how the parts of the mechanism would have been made and put together. He asked the museum staff if anyone there knew about making geared mechanisms. The answer came without hesitation: 'Michael Wright.'

And so Bromley ended up in Wright's office. He came often, whenever he was over from Sydney, chatting to Wright over his sandwiches and learning about the practicalities of nineteenth-century planing machines and treadle lathes. He put together a proposal, which he handed to the museum's senior curator of computing, Doron Swade, in May 1985. It would be one of the most ambitious scientific reconstructions ever attempted, likely to cost at least a quarter of a million pounds. But persuasion was one of Bromley's talents. Under Swade's supervision and with the help of some industry sponsorship, Difference Engine No. 2 was duly built.

It carried out its first calculation on 29 November 1991, a month short of the bicentenary.

Over their lunches, however, Bromley and Wright didn't just talk about Babbage. They covered the whole world of mechanical marvels and out of their conversations grew friendship. Wright told Bromley about his growing interest in the Antikythera mechanism and his dream of going to Athens to study it. He showed him Price's papers, explained where he could see that Price had gone wrong, and the two swapped ideas about how the device might have worked.

The Antikythera mechanism piqued Bromley's interest immediately. Babbage's designs belonged to the line of digital calculators and computers, in which calculations are converted into numerical equations and the answer is given as a string of digits. This idea is familiar to us now because it's the method used by modern electronic computers. But the Antikythera mechanism is part of a tradition of analogue computing,[2] in which problems are modelled more directly and the output is continuous, for instance as a dial on a

2 Is it correct to describe the Antikythera mechanism as a computer? The term was first applied by Derek Price when he called it a calendar computer, but some scholars have suggested that the word 'calculator' would be more appropriate. This book follows the guidance of Doron Swade, a computer historian who until recently was senior curator of computing at the Science Museum in London. He argues that it is short-sighted to restrict the term 'computer' purely to the programmable electronic devices that we are familiar with today. In his view, the computing tradition started much earlier. He defines a computer as any device that can not only calculate a mathematical function, but also display the answer on a numerical scale. So a spherical model of the solar system would not count as a computer, for example, but the Antikythera mechanism most certainly does.

scale. If you're trying to solve a problem of trigonometry, for example, the digital method would be to derive the appropriate equation and work out the answer on a pocket calculator. But you could instead make a scale drawing of the triangle in question and simply measure the answer. That would be the analogue approach.

A slide rule is a simple example of an analogue computer, as is the astrolabe. Slightly more complex were the mechanical gun-aimers used during the Second World War. They had two metal arms to represent the angle above the horizon and the distance to the target aircraft, so once they had been set appropriately you could read off the height and horizontal distance to the target.

The first programmable analogue computer was the Totalisator – an Australian invention first installed at Newcastle Racecourse in New South Wales in 1913. It used banks of differential gears to calculate the amount of money to be paid to winning punters from a pool. Bromley had one of the earliest models in his garage, but he had never heard of anything like this strange Greek clockwork device. This mysterious machine represented the very beginning of the computing tradition, digital or analogue – it was the first known example of an object that people had built to think for them, to work through mathematical equations and display the answer on a numerical scale. Instantly, Bromley was mentally converting its gear trains into circuit diagrams and a new plan began to form in his mind. He would be the man to solve the Antikythera mechanism.

In particular, Bromley had a theory that the device couldn't have been driven by the slow-moving big wheel – there just wouldn't have been enough power to drive all the subsequent gear trains. He tweaked the gears and came up with his own variant, which was driven by the much faster moving turntable – the same one that carried Price's differential gear. Back in Sydney, he tried out a rough model of the mechanism using Meccano gears, then worked with a clockmaker called Frank Percival to build a proper reconstruction. They had terrible trouble getting the sharp, triangular teeth to mesh properly, but after rounding them off at the edges it worked sweetly – much better than Price's model. Wright, meanwhile, was rapidly losing faith that any part of Price's reconstruction could be trusted. The only way to find out for sure was to go and study the fragments.

Then, just before Christmas 1989 Bromley swept into Wright's office bearing a triumphant air not dissimilar to the one that Judith Field had worn when she brought him the Byzantine sundial six years earlier.

'I've just come from Athens!' he announced grandly. 'The museum has given me permission to work on the Antikythera mechanism!'

Wright's jaw dropped. All this time he had been dreaming about studying the mechanism and now Bromley – his friend, the man he had confided in – had stolen his idea. There was an unwritten rule with Greek antiquities that when access to an artefact was granted to one researcher it was withheld from all others until that person had published their results. Wright had been shut out. Not for the first time he felt sorely jealous of these university academics who got to swan around from institute to institute studying whatever they chose. He knew his stuff, and he was better equipped to tackle the Antikythera mechanism than this man was. If only he could get the chance to prove it.

Bromley was heading back to Athens to examine the fragments in just a few weeks' time and Wright realised there was only one way that he could get there, too. He swallowed his pride and asked Bromley to take him along as his assistant.

Bromley agreed, so Wright, who had never organised a foreign trip in his life, collected his thoughts and his papers and booked some time off. The two of them flew to Athens in January, arriving late on a Sunday night – Bromley stayed in a comfy B&B across town from the museum, while Wright managed to snatch a last-minute bed at a nearby student hostel.

The next morning the pair met at the museum, where they were greeted by Petros Kalligas, seasoned curator of the museum's bronze collection, who had worked with Price and Karakalos back in the 1970s. He was a charming old man and spoke excellent English, which came as a relief to Wright, who hadn't realised until he arrived in Athens that modern Greek was somewhat different to the ancient language he had learned at school.

After treating the pair to a shot of stiff black coffee, Kalligas led them to a table in the corner of the bronze conservation workshop, surrounded by pieces of old statues.

On the table, lying on sheets of tissue paper in a tray, were the Antikythera fragments.

Wright's first thought was how small they were – photos hadn't prepared him for how tiny the instrument was, each piece dense with intricate detail. His second, jubilant, thought was: 'Price didn't notice half of this!' Eagerly, he pulled on his white cotton gloves – standard issue for museum curators when handling delicate artefacts – but Kalligas gently stopped him. 'We find that much less of the material rubs off on bare skin,' he said. And he left them to it.

Wright and Bromley spent the next month carefully measuring, photographing and noting down every detail of the fragments and checking their observations against those recorded by Price. They worked every day until the museum staff ushered them out in mid-afternoon. Then there was food and some light relief, wandering the sights of Athens or settling in to one of the many little backstreet bars, where they drank home-made wine from tin jugs and ate *mezedes* from a tray as they discussed the finds of the day. After that it was generally back to work, but sometimes they'd move from wine to ouzo as the afternoon turned to evening and then the early hours of the morning. A particular favourite was To Gerani (The Geranium), a down-at-heel drinking den on Tripod Street.

One day early in their stay Bromley was unwell after a particularly lively night out. Wright, who had stayed in, went to the museum alone, watched over only by a bored-looking assistant called Tassos and a few buzzing flies. The pieces looked like green, flaky pastry, he thought, and felt like it too, only a bit heavier. He touched them as lightly as possible to minimise the amount of dust that crumbled away at the slightest contact with his fingers. These fragments had survived so long, through the vicissitudes of 2,000 years of history, yet they were now so fragile. Every tiny piece might contain some vital piece of information that if he wasn't careful could be lost forever. One of the fragments consisted of a round, corroded mass with a thinner, curved piece of the zodiac scale sticking out to one side. He turned it over to see what was on the back, but as he did so, he heard a sickening click and the scale broke into two before his eyes.

Wright was distraught. He had been looking forward to this opportunity for years and now he was finally able to study the fragments for himself he had broken one of them! Surely he would be banished from the museum forever for his clumsiness. He rushed past Tassos, out of the door and down the corridor into Dr Kalligas's office.

'There's been an accident,' he said, his voice shaking. The curator looked up from his papers. 'An accident? What sort of accident?'

'One of the pieces has . . . has . . .' He forced out the word. 'Broken.' 'Hmph!' said Kalligas. He strode out of his office and into the workroom, Wright following nervously behind. He examined the broken pieces. 'Hmph!' he said again. Then he was silent for a few moments, until, finally, he spoke.

'It happens,' he said. 'It happens to everyone. It happens to Tassos. It even happens to me. Now it has happened to you. Go home, have a drink, get some sleep. Then come back tomorrow.'

The next day, Wright returned with Bromley to survey the damage. It wasn't surprising that the fragment had broken. Inside, there was hardly any metal left – just a tiny sliver of pink remained within the flaky green. The two of them got a couple of photos of the break. Then Tassos stuck the pieces back together with superglue.

As Wright and Bromley's work progressed, it became clear that as well as missing a lot of the detail, Price had been mistaken in several important respects. For example, the fragments didn't fit together as Price had said they did. His positioning of fragment D – the lonely cogwheel – in a back corner had been a key part of his reconstruction of the gear train leading to the upper back dial. Without it there was no evidence whatsoever for Price's idea that it had been a four-year dial. Several other gearwheels, too, were not where Price had put them in his reconstruction. They would have to throw out much of his model and start again.

There was also a new piece of the mechanism – called fragment E – which Kalligas had found in a back room in 1976, too late for Price's study. It was just a few centimetres across and formed part of the lower back dial, which Price had thought displayed the phase of the Moon, as calculated by the differential gear.

After photographing the fragments and noting down every

visible detail, Wright and Bromley moved on to radiography. There was an X-ray machine in the lab of the museum's scientist, a bright, matronly woman called Eleni Magkou. Charalambos Karakalos still guarded his images jealously, so they decided to X-ray every piece of the mechanism again. Magkou left it to one of her technicians, Giorgos, to do the work. But the results were puzzling: the films were coming back fogged and with a distinctly yellow tint.

Their time up, the pair left Athens disappointed with their X-ray images and grappling with the same problem that had stumped Price. With so many layers of gearwheels all appearing on top of each other, it was impossible to tell how the mechanism was arranged. To get any further they would need to separate out the different depths.

Soon after they returned to London, Bromley gave a lecture on the Antikythera mechanism to the Antiquarian Horology Society. It annoyed Wright, because Bromley kept referring to the project as if it were solely his, but one useful thing came out of it. A member of the audience called Alan Partridge came up to them afterwards. He was a Meccano enthusiast (he had a room at home full of the stuff) and like Bromley he had used it to model Price's reconstruction of the Antikythera mechanism. Partridge was a retired doctor and he suggested that they build a crude linear tomography machine. He had worked in hospitals in poor countries such as Nigeria, so he knew how to do things low-tech.

Linear tomography was first developed in the 1920s and was used during the Second World War to locate bullets and shrapnel in the bodies of wounded soldiers. You lie the patient on a couch, the X-ray source on one side and the film on the other. During the exposure you keep the patient still, while you slide the source and the film together, so that only one plane within the patient stays in focus, while all the others blur. By adjusting the distance between the source, the patient and the film, you can take a series of images, each with a different plane in focus, like imaging a series of slices through the patient's body.

Wright decided to build the necessary equipment in his workshop at home. The X-ray source at the Athens museum was going to be too large and heavy to move around. But he could get the same effect another way. He constructed a cradle from aluminium castings and plywood that would hold the fragment and the film at a set distance apart. Once fixed inside they could both be swung together, while the X-ray source stayed still, giving the same effect as conventional tomography.

As well as building the cradle, he read up on the theory of tomography, spending his evenings studying tables of exposure times. And he made a fake fragment to test out his skills by pouring casting resin over some old cogwheels and metal plates from his scrap box so that they set in a lump. The equipment worked beautifully and the resulting images allowed him to separate out depths of less than a tenth of a millimetre, surely enough to resolve even the tiniest detail inside the Antikythera mechanism.

The next winter the pair returned to Athens with Wright's

tomography cradle packed in a suitcase, and armed with boxes and boxes of X-ray film donated by a kind gentleman at Agfa. But before they could start on their tomography, they had to work out why the quality of the X-ray images had been so poor.

Eventually they realised that Magkou's lab had no money to pay for developing chemicals. So rather than make up his own, Giorgos, the technician, would wait until the museum's unsuspecting photographer went on his lunch break, then use his developing baths. The silver grains from the X-ray film must have played havoc with the poor man's photos. Bromley soon sweet-talked Sydney University into providing some extra funds and picked up the necessary supplies, much to the astonishment of Eleni Magkou. Bromley had spent more on chemicals than her entire consumables budget for the year.

The images were better, but still not right. Then Wright got permission to 'help' Giorgos in the darkroom and it soon became clear that his attitude was relaxed, to say the least. He didn't believe in watches and would develop his plates according to the time it took him to smoke a cigarette; its end glowing orange in the otherwise inky black darkroom. Luckily he was only too pleased to wait outside while Wright took over.

With the technical difficulties solved, they worked every moment they could with Bromley taking the exposures and Wright doing the delicate job of developing the plates – spending hours on end in the darkness before emerging blinking into the bright Athens sunlight. Then a smiling Bromley would drag him off to the nearest bar for a glass or three of retsina.

They continued with this routine every winter, Bromley coming during his university's summer vacation and Wright using his precious holiday time. Finally, three years and more than 700 exposures later, they were done. It was February 1994. Wright was confident that he might be limited by the sorry state of the fragments, but not by the quality of the radiography. Within their piles of images he knew that the answer to the Antikythera mystery – if there was an answer – lay waiting to be discovered.

Then Bromley dropped a bombshell. He thanked Wright for his work, but announced that as the lead partner on the research project, he would be taking all of the radiographs back to Sydney. The best way to study the images was to scan them into a computer and he had a student who was waiting to get started.

Wright, once again, was horrified by the behaviour of the man he had thought was his friend. This was not fair play. For the last five years he had spent every spare moment thinking, planning, preparing for his work on the Antikythera mechanism. He had built his own equipment, learned new skills and patiently coaxed details out of the stubborn fragments that no one else could have hoped to glean. Now Bromley was waltzing off to the other side of the world with his precious images.

But Wright was exhausted and didn't have the energy to

argue. He hated confrontation and felt he had no chance of winning against the forceful, supremely confident Bromley. Wright had given the project everything he had. He flew back home with nothing.

7

Mechanic's Workshop

The heavens themselves, the planets, and this centre,
Observe degree, priority, and place,
Insisture, course, proportion, season, form,
Office, and custom, in all line of order.

— WILLIAM SHAKESPEARE, *TROILUS AND CRESSIDA* (I. 3)

BING! THE LIGHT above Wright's seat instructed him to fasten his seatbelt for arrival in Sydney. His mind chased in uneasy circles, and he felt slightly sick. A stewardess was looking at him and speaking; with some effort he tuned in to discover that she was asking whether he'd had a pleasant flight. As ever, he gave an honest answer.

'You did your best,' he said. 'But I'm terrified at what I'm going to find when we land.'

The years following Wright's last trip to Athens with Bromley had been – not to put too fine a point on it – bleak. The pressures of work and life had taken their toll. He was separated from his wife and children and living in lodgings; separated from his workshop, too, with his tools mostly in storage (although his landlord had been kind enough to let him put his lathe in the scullery). And at the Science Museum his bosses were pushing him to take time off to deal with his depression, a move which he was convinced was all part of a plan to sack him.

Then, just as he had begun to set up home on his own, he fell while decorating the bathroom and punched his hand through the china lavatory, cutting through blood vessels, tendons and a nerve at the wrist. He was told he would never use the hand again – a devastating prognosis, which fortunately turned out to be inaccurate. Nevertheless, it took months of therapy to learn how to cope with the crippling loss of sensation and awkward movement he was left with, and years to regain confidence in the use of his hand.

Meanwhile, any correspondence from Bromley had dried up. Wright had never seen his precious images again, and Bromley's promise to digitise the radiographs came to nothing. At first, Bromley had sent him odd snippets of information to see what he made of them, as if deliberately teasing him, but then months went by and Wright heard nothing.

So much of human life is wasted doing pointless things, he thought. So much of what we do is futile. For the last ten years of his life solving the Antikythera mechanism had been the one thing that seemed worth doing; his one chance to make an important and lasting addition to human knowledge. There were a lot of things he wasn't good at – just looking around at his life at the time made that perfectly clear. But this was the one challenge in which he had the skills needed to succeed where no one else could. Without the Antikythera mechanism, Wright wasn't sure what he was doing on the planet at all.

He kept mulling over a detail that he had seen in Athens – a fundamental problem at the heart of Price's reconstruction, casting his whole model in doubt. Price had interpreted a train of gears leading from the main drive wheel back to a wheel centred around the same point as encoding the Metonic 19-year cycle – this was the part that converted the once-yearly motion of the main wheel into the speed of the Moon around the zodiac. The two motions ended up going in opposite directions, then both fed into the differential gear. By subtracting one from another, the differential gear supposedly calculated the phase of the Moon.

Wright had seen that there was an extra wheel at the end of this train, which Price had missed. It had the same number of teeth as the one before, so the rate of rotation of the lunar pointer wasn't changed. But the direction of the motion was. This made sense, because the train could then drive the Sun and Moon pointers in the same direction around the zodiac dial. (Price had been forced to imagine another big wheel the same size as the main drive wheel, driven by the other side of the crown wheel and thus in the opposite direction, to drive the Sun pointer.) But it caused a huge problem further down the line. If the rotations of the Sun and Moon were fed into the differential gear going in the same direction, then they wouldn't be subtracted, they would be added. And adding the Sun to the Moon makes no sense at all. Something was very wrong.

But without his radiographs Wright was left with nothing but questions and no way to answer them. He cursed Bromley for his betrayal and he cursed himself for not being stronger and standing up to him.

Wright had no way of knowing that on the other side of the world the research project was not going as Bromley had hoped. In fact, Bromley was desperately trying to hide from Wright – and everyone around him – that he was ill and increasingly unable to work. Then, as the end of the century approached, Wright received a letter from Bromley's wife explaining that her husband had been suffering for years from a form of cancer called Hodgkin's lymphoma and that he was deteriorating fast. *If you want to see him*, she wrote, *you have to come soon.*

Wright was desperate to retrieve his radiographs, but he felt he couldn't go unless Bromley asked him. He was frustrated with Bromley, had even hated him at times, but to visit uninvited would be like saying to his friend that he knew he was going to die. Bromley finally wrote to Wright in 2000 telling him that he had been given just six months to live. That was enough. In November, Wright summoned up his courage and flew to Sydney.

He turned up on Bromley's doorstep exhausted and nervous; it was nearly six years since they had last seen each other. But when his friend opened the door, Wright's fear turned to shock. Bromley was barely recognisable. Wright was used to seeing him full of energy and the centre of attention, seemingly capable of achieving anything and of persuading others to come along for the ride. Now his beard and smile were gone and his round, bonny face had become chalky-white, the skin drawn tight over bone.

Wright stayed for three weeks in the back room of Bromley's bungalow, surrounded by the radiographs he had been separated from for so long. Some days Bromley never made it out of bed, so Wright left the house and toured Sydney's museums. But when Bromley was well enough, the two of them sat and talked for hours about their time in Athens, about the Antikythera mechanism, and about what went wrong.

Bromley was tired, weak and depressed. After years of denial about his illness, he couldn't escape the fact that he was dying. He told Wright in one of his more lucid moments that for years he had dreamed that his name – and his name alone – would be attached to the final solution of the Antikythera mechanism; that was why he had guarded the results so jealously. To Wright it felt like the closest that Bromley was going to come to a confession; if not an apology, then at least an admission that he had treated Wright badly.

Still, trying to persuade Bromley to hand over the radiographs before he died was a horrible task.

To Bromley, knowledge really was power. Throughout his life he had defined himself by what he knew and others didn't. It was what drove him to become the world's foremost expert on Charles Babbage, and it was what drove him to return to Athens year after year to study the Antikythera mechanism. Knowledge wasn't to be shared, it was to be held onto like a currency, to be bartered at a later date. Once, on one of his trips to London, he gave a seminar on the Antikythera mechanism to an audience of curators at the Science Museum. At the end, as is customary at such events, one of them raised his hand and politely asked a question. Bromley looked straight at his inquisitor, eyes twinkling, the corners of his mouth curling through his beard into the hint of a smile.

'That,' he said finally, 'is for me to know and you to find out.'

So for Bromley to give up all of his data to Wright would be the ultimate defeat, it would be an admission that all hope was gone. Wright was torn by guilt; he had to make Bromley face the fact that whatever his dreams had been, he would never be the one to solve the Antikythera mystery. But he could see the pain in the lost, shrunken man before him.

Some days, Wright thought he had won.

'This is a hopeless case,' Bromley would announce. 'Take the data if you want.' But on other days he'd change his mind, saying that Wright had convinced him the project was worth pursuing: 'When I'm stronger I'll have another go.' Eventually, Bromley's wife found a way.

'You've done this great body of work,' she coaxed. 'Let Michael go and make sense of it all, so that your efforts can be recognised.'

Bromley did keep some of the material – most of the

photos and the clearest radiographs – but he handed over the rest to Wright at the end of his stay; a last gift to the living from the soon-to-be dead.

When it was time to leave, Bromley insisted on driving Wright to the airport, though the effort drained the air from his lungs and the blood from his face. They said goodbye on the forecourt.

'It's been good knowing you.'

'Have a safe trip.'

It was the last time they saw each other. Bromley lasted longer than the six months the doctor gave him, but he finally succumbed in September 2002, without making any further progress on the Antikythera mechanism. He was 55.

Wright was asked to write an obituary for Bromley. Unlike the perfectly glowing accounts that appeared elsewhere, his article was the painful result of an urge to tell what he felt was the unvarnished truth. 'If I sometimes resented the way in which Allan took, and kept, control of the project, I recognise that without him I might well never have got to Athens at all,' he said. He ended with four simple words: 'I will miss him.'

After Wright returned to London with the radiographs, still working at evenings and weekends, he developed an idea that had been forming in his mind since he first saw the fragments in Athens. It was the possibility that on the front of the mechanism there had once been many more gears, which modelled the movements of the planets.

It was a bold idea, but several lines of evidence were

leading him to the same conclusion. When Wright had examined the fragments directly he saw the remains of brackets sticking forwards from the large spoked wheel. It looked as if they had once carried something round on this gear as it turned. Price had noted these brackets, but ultimately ignored them. In his reconstruction there wasn't room for any extra mechanism here, because his second big wheel – his Sun wheel – had turned directly in front. But Wright now knew that there was no need for a Sun wheel. The way was clear for extra structures. What were they?

Wright suspected there were more gearwheels. He had seen wheels carried round on other wheels before – it was a common device in the astronomical clocks he knew from the Science Museum, particularly for calculating the movements of the planets. A train of gears mounted on a turntable would drive a gearwheel around at a certain speed, at the same time as it was being carried round in a bigger circle. Whereas a differential gear has two inputs and one output, here there is just one input and one output, but the output isn't constant – it speeds up and slows down relative to the central axis as the little wheel turns. This type of gearing is still called epicyclic or planetary gearing, although it's now more likely to be found in vehicles or industrial machinery than in astronomical displays.

Unlike the Sun and the Moon, the planets don't follow smooth paths across the sky. They change speed, stop and zigzag about, so much so that their name (from the Greek word *planetes*) means 'wanderer' or 'vagabond'. These erratic movements in the heavens upset the Greeks of classical times, because they liked to think of everything in the universe as perfect, and perfect motion was uniform motion in a circle. The structure of the universe reflected the nature of the gods, so there could be no question of any sort of deviation or irregularity. Reconciling the wandering motions of the planets with this idea of perfect circles became one of the most pressing philosophical problems of the day.

In the fourth century BC one of Plato's students in Athens, called Eudoxus, came up with a system of concentric spheres. Those that carried the planets slid over others, all rotating in different directions, with the Earth in the middle. The effect was that each planet traced a sort of figure-of-eight curve. It was a cunning idea, but it didn't match the actual movements of the planets very well. Then in the third century BC a mathematician working in Alexandria called Apollonius developed a much better idea: epicycles. He imagined the planets looping the loop – in other words, travelling in a small circle at the same time as the centre of that circle moved around the Earth.

This explained why the planets appeared to speed up and slow down, and why sometimes it even looked as if they were going backwards.

The concept works because it does actually bear some relation to reality. As we observe a planet, it is circling the Sun at the same time as Earth is, so the motion we see is the combination of those two circles – the planet's orbit and ours. When we look at Mercury or Venus, which are closer to the Sun than we are, the motion we see is a combination of our orbit around the Sun (this is the bigger circle) and the planet's orbit around the Sun (the smaller circle). When we look at the planets that are further from the Sun than us – Mars, Jupiter and Saturn – we're the ones looping the loop. Our path around the Sun is superimposed on the bigger circle that is the planet's orbit around the Sun. Apollonius didn't know any of this, of course. He was just trying to come up with a geometrical model that could explain the way that the planets looked from Earth.

It's quite straightforward to translate this model into the language of gears. If you want to model the motion of Venus, for example, you need a big turntable that rotates at the speed of the Earth around the Sun (or the Sun around the Earth, from a geocentric point of view). Then you need a smaller gear that spins around on that turntable, to simulate Venus's orbit around the Sun. The size and speed of this smaller gear relative to the big turntable are determined by the size and speed of Venus's orbit compared to Earth's. If you imagine a pin sticking up from a point on the edge of the smaller wheel, then the speed with which the pin circles the big wheel's central shaft represents the motion of Venus as we see it. The astronomical clocks of Renaissance times used slotted levers to translate this motion back to a pointer on the zodiac dial. The dial pointer would be driven by a shaft that was itself turned by a lever that had a slot cut into the end of it, so it looked a bit like a tuning fork. This slot fitted over the pin on the epicycle wheel. As the epicycle wheel looped the loop, the pin slid up and down in the slot, driving the lever, and therefore the pointer, around at varying speed.

All of the necessary sizes and speeds of the cycles and epicyles can be determined from straightforward observations of the planets' movements. Ptolemy, working in Alexandria in the second century AD, worked out the appropriate mathematical equations. But other Greek astronomers were certainly thinking along similar lines before him, even if their figures haven't survived.

So could the Antikythera mechanism have included epicyclic gearing to model the motions of some or all of the planets? The inscriptions that Price had originally noted (but passed over in *Gears from the Greeks*) were a hint in this direction – Venus was mentioned, and there were several mentions of 'stationary points', the moment at which a planet appears to stop and change direction. But that wasn't all. Wright saw that protruding forwards out of the centre of the big four-spoked wheel was a sturdy square pipe that was fixed on to the metal plate behind. The square shape suggested that some wheel must have sat here, with a square hole to make sure that it didn't slip around the central pipe. This is just what you would expect if there was epicyclic gearing riding on the big wheel – a fixed wheel in the centre would drive an epicyclic gear train as it was carried around on its turntable. This idea also explained the size of the four-spoked wheel – it needed to be so big because it was carrying other gearing.

The four-spoked wheel moved around at the speed of the Sun around the Earth. That would be perfect for carrying gearing to model the two inferior planets, Venus and Mercury. But ancient Greek astronomers saw all of the planets as equally important. Wright felt sure that if the designer of the Antikythera mechanism had modelled Mercury and Venus, he would have included the others as well. This was a bit more complicated, because it would require a separate turntable rotating at the appropriate speed for each planet, with the epicycles then rotating at the speed of the Sun. But it could be done using just the same mechanical techniques. During 2001 Wright presented his ideas at a conference in Olympia, Greece, taking along a little cardboard model to demonstrate how the planetary gearing might have worked.

Wright knew that suggesting so much extra gearing for which no trace remained was likely to be controversial. And although the epicyclic gearing he was proposing was no more complicated than the differential gear that Price had suggested, he was sure that he'd face strong scepticism about whether the Greeks would have been capable of such a thing. It was a step up from simply modelling a body going around the Earth at a constant speed. This was taking the latest mathematical theory about variations in the motions of the planets and translating it into mechanical form.

So Wright decided to do what he did best. He would make a model of the mechanism by hand, using traditional materials and methods, to prove that the Greeks could have done it. As soon as he got back from Olympia he gathered some pieces of scrap metal – including the name plate from an office door and a pub door kicking plate – and started to make his own Antikythera mechanism. By this time he had married again – to an understanding woman called Anne who worked across the road at the Victoria and Albert Museum – and he'd made a new workshop in one of the rooms of their period home in Hammersmith, pushing back the tide of books until every inch of floor, wall, shelf and bench space was covered with tools and old metal gadgets and instruments, from replicas of ancient astrolabes to twentieth-century trombones.

By the end of the year he had reconstructed the front of the mechanism: the gears calculating the motions of the Sun and Moon, as well as epicyclic wheels for Venus and Mercury. He now needed to construct separate turntables for Mars, Saturn and Jupiter. And he figured that if the mechanism contained epicyclic gearing for the planets, it would probably have done so for the Sun and Moon as well.

Only one orbit is involved in the apparent motions of each of these two bodies from Earth – us around the Sun, and the Moon around us. But the Greeks knew from their observations that this couldn't be the whole story – both the Sun and Moon appear closer at certain times than others, and their speeds vary in a regular pattern. This is because the orbits of the Earth and Moon, like those of the planets, are not regular circles, but elongated ellipses with the body that they are orbiting around located nearer to one end.

The Moon, for example, spends part of its orbit relatively close to us (so it looks larger in the sky and appears to be moving faster than average) before it swings off around the more distant tip.

The Greeks were so committed to the idea of the heavens consisting of perfect spheres and circles that they would have found it impossible to contemplate such a thing. Instead they explained the variations using different combinations of circles – either epicycles or what's called an 'eccentric' model. This assumes that the Moon or Sun is moving in a perfect circle around the Earth, but that the centre of its orbit is slightly offset from the Earth. So one end of its orbit will be slightly closer to us and the other end will be slightly further away. Because the orbits concerned aren't very elongated this actually works quite well as an approximation to reality. Wright added two more epicyclic gear trains to his model so that the pointers on the zodiac dial incorporated the varying motions of the Sun and Moon. The Sun epicycle sat on the main wheel along with those of Venus and Mercury, while the Moon epicycle needed its own turntable. He kept the old Sun pointer, however, which was simply going around at the Sun's average speed, because this showed the date against the calendar scale.

Everything was falling into place. The epicyclic gearing fitted into the mechanism so naturally he knew he was right. But as he worked on the planetary display, the back of the mechanism – Price's supposed differential gear and the function of the back dials – remained a mystery.

Then Wright heard that he had competition.

A film-maker called Tony Freeth, who lived down the road in Ealing, west London, was apparently campaigning to persuade the Athens National Archaeological Museum to let him image the Antikythera fragments again, using the latest X-ray technology. Freeth was working with Mike Edmunds, an astronomer from Cardiff University, and a team of prestigious Greek scientists.

Wright had come across Edmunds before. Judith Field and Edmunds had studied together at Cambridge University and one day she got a call from him. Edmunds asked her about the Antikythera mechanism, so she gave him Wright's number.

If he calls, you must tell him all you can,' she told Wright. 'He's a serious astronomer, a professor! Make sure you give him a good answer.'

So when Edmunds called, Wright spoke to him at length, telling him all about the problems he saw with Price's reconstruction, how the mechanism may have displayed the movements of the planets, and what he thought needed to be done next. Edmunds told Wright that a research student of his, Philip Morgan, was engaged in a project to reassess the mechanism and that at the end of it, they planned to write about their work.

'I'll let you know when it comes out,' he said.

Wright saw the article in print at the beginning of 2001. To most eyes it would have been a fascinating account of a little-known ancient mystery, but Wright was horrified.

Much of it was a review of Price's work, which, however well-researched, he felt should be discounted. As far as he was concerned, at least, the most original and worthwhile points in it were the ideas that he had shared with Edmunds on the phone – the possibility that the device had been a planetarium, and his ideas for future research. At the end of the paper, Edmunds and Morgan acknowledged a 'communication from M. T. Wright', but that was it. There was no mention of all the years he had worked on the mechanism or the progress he had made. A familiar feeling crept through him like icicles. Once again, he had shared his ideas in good faith. And once again, he felt he was being sidelined.

Straightaway, he wrote a letter of complaint to the journal and it was published a couple of issues later. 'Readers may wish to know about the subsequent and continuing work by Bromley and myself,' Wright said 'which will oblige all who are interested in this mechanism to trust less implicitly in what Price wrote.' After outlining some of the work he had done so far, he finished with barely contained frustration: 'It is surprising that Edmunds and Morgan do not mention our work, since the communication that they acknowledge was a telephone conversation between Prof. Edmunds and myself in which I outlined it.'

Beneath the letter, the journal ran a reply from Edmunds – an equally thinly veiled dig at Wright's lack of progress so far. So when Wright heard that Tony Freeth and Mike Edmunds were working to get access to the Antikythera fragments, just as he was finally starting to get somewhere, he felt dismayed and angry. Later, when Freeth approached him to ask if he would support their campaign, he sent a frosty reply, explaining that as an employee of the Science Museum it would not be appropriate for him to tell staff at another museum what to do. Besides, there was no need for new images. His own radiographs were perfectly adequate and he knew he could solve the puzzle, if only everyone would leave him alone long enough to do so.

For the time being, the Athens museum had turned Freeth and Edmunds down, because Wright was still at work on his data. But with the number of eminent astronomers on their team Wright feared it would not be long before they succeeded in imaging the fragments, and if they did so before he solved the mechanism, all of his years of effort would have been in vain. He had to get his work published, and fast.

After 500 hours work – exactly 100 days of weekend working and sleepless nights – Wright finished the planetary display of his reconstruction, and published it in a specialist clockmakers' journal in May 2002.

But things at the Science Museum soon went from bad to worse. As part of the ongoing process of modernisation, a team of management consultants had been called in to streamline the way the institution was run. All of the curators had to apply for new jobs, and it was clear that academic research would not be a priority. The museum was under pressure to justify the money it was spending by reaching out to the public and getting as many people as possible in through its doors. That meant thinking up new ways to make science glamorous and exciting, with interactive exhibits, snappy soundbites and multimedia displays. Old-fashioned curators hunched over dusty instruments in glass cases were not part of the vision.

To be interviewed as if he were a stranger after more than 30 years of service at the museum was not something that Wright felt able to swallow. When asked to give presentations by his interview panels in early 2003, he lectured them on 'How to Run a Museum' and 'The Importance of Saying What You Mean'. Still, the redundancy package gave him enough to live on for a while, and now he had more time to work on the Antikythera mechanism.

By this time Bromley had passed away and his widow had sent Wright the remaining images that she could find. So he began trying to solve the rest of the device, with nearly 700 radiographs to work from. He couldn't see any way to publish all the images, and even if he could they wouldn't make any sense to anyone unless he could come up with a suggested reconstruction. Although his tomography cradle had ensured that in each image only features in the desired plane were sharp, the others were all still there as blurred grey streaks, so interpreting the details was a time-consuming and specialised task. He studied the radiographs over a lightbox one by one with a magnifier, as Emily Karakalos had done, until his eyes streamed and his head hurt.

By this time, however, Wright's son Gabriel was studying for a doctorate in medical imaging at Oxford University.

Gabriel's lab had the necessary equipment to scan in the radiographs as high-resolution digital files. He patiently scanned in all 700 radiographs and set them up on his father's computer with some basic image manipulation software. By the end of 2003 Wright could zoom in on his images, alter their brightness and contrast, and flick effortlessly from one image to another. Before he had been estimating tooth counts by cutting out transparent circles of different sizes and divisions and laying them over the film to compare with the wheels beneath, but now he could measure them accurately with the click of a mouse. He found himself agreeing closely with Karakalos's counts, and not with the numbers accepted by Price.

Things were really starting to move now and he published a string of papers, one for each step he made, all in specialist publications that dealt with clockwork and scientific instruments. When he measured the dials on the back of the device, he realised that although the rings were concentric, the two halves of each dial were drawn around a different centre. In other words, each dial was not made up of several separate rings, but a single spiral. The upper spiral had five turns and by measuring the marks on it he calculated that each revolution of the pointer represented 47 divisions, making 235 in all. He realised that the spiral must have displayed the 235 months of the Metonic 19-year cycle, as calculated by the gear train under the front dial.

And when he looked at the little dial next to this spiral, Wright saw that it was divided into four. The Greeks had used another period, called the Callippic period, made up of four Metonic periods so that it was 4 x 19 = 76 years long. The year was known to be 365 1/4 days long, and this longer period got rid of the awkward quarter days. This period was even mentioned in one of the fragmentary inscriptions from the mechanism, so it made sense that it would have been shown on one of the dials.

The gear train leading to this dial was lost, but by adding in three extra wheels Wright came up with a simple train in which the little dial turned once in every 20 turns of the main pointer, so that it would show where you were in the Callippic cycle. It could have been used to help keep track of long intervals of time when the mechanism was being wound forwards or backwards, and would also have been useful for converting dates given in the Egyptian solar calendar (as displayed on the front dial) with any of the various local lunar calendars.

On the front of the mechanism Wright also made sense of a strange circular arrangement that seemed to be stuck on to the front dial. Price had seen it and thought it might be the remains of a folding crank handle, but from knowledge of later astronomical clocks Wright recognised it as a Moon phase display. In his radiographs he could clearly see the remains of a little wheel at right angles to the others, designed to pick up the relative motion between the Moon pointer and the Sun pointer, to turn a little Moon ball on its axis. There was a perfectly round hollow in the crumbling fragment, showing that the ball itself had not corroded and had fallen out only after everything around it had done so. Wright suggested that it might have been made of ivory, with one half blackened with ink. The ball was set into the central boss of the Moon pointer so that only the front half showed, and it would have spun in time with the phase of the Moon, showing all black when the Moon wasn't visible, through a sliver of light when the new crescent appeared, to a whole white face at full Moon. Wright was stunned to see it in such an ancient machine.

Then there was the problem of Price's differential gear. Price had described two linked inputs that sat on a turntable, driving it around at a speed that was half the sum of their two rotations. But when Wright looked closer, he could see only one input. So it wasn't a differential gear at all, but appeared to be an epicyclic one, similar to the ones he had already suggested for the planets. It wasn't in the right place to be modelling any of the planets they would need to be on a turntable concentric with the front dial. But there is another use for epicyclic gearing: to calculate gear ratios that are too complicated to achieve by normal fixed gears. It was commonly used for this purpose in the elaborate astronomical clocks of Renaissance Europe.

Wright drew up spreadsheets of all the possible numbers of teeth for the gears in the train, but couldn't see what it was meant to calculate. And he noticed a couple of other strange features that were hard to explain. The first was that the turntable had 223 teeth around its edge, which didn't seem to engage with anything. That was odd – 223 is a prime number and you'd only bother to make such a wheel if you needed the prime number for a particular gear ratio. It didn't make sense to cut such a wheel and then just use it for a turntable that didn't need any teeth at all.

And on the turntable he saw a double wheel system in which one small wheel sat almost, but not quite, on top of the other. The bottom wheel had a pin sticking up from it, which engaged with a slot in the top one, so that as the wheel with the pin rotated it drove the other wheel around. Because the wheels rotated around slightly different centres, the pin from the bottom wheel would slide up and down in the slot, towards and away from the centre of the upper wheel, introducing a wobble into its speed of rotation.

Wright had seen such mechanisms in astronomical clocks. They were used to model the fact that planetary orbits are ellipses, not perfect circles, so their apparent speed varies. No one had worked out the maths to model the planets that way at the time the Antikythera mechanism was made. But the ancient Greek astronomer Hipparchus did use such a wobble in equations to describe the motion of the Sun and Moon.

It was a wonderful discovery – the earliest example of such a pin-and-slot mechanism by close to 1,500 years. And it gave valuable support to Wright's idea that pins and slotted levers were used in epicyclic gearing at the front of the device to model variations in the motion of the Sun, Moon and planets. But positioned as it was towards the back of the mechanism and on a mysterious turntable, he was stumped about the purpose of this particular pin-and-slot. Although he realised the similarity to Hipparchus's theory, he couldn't see how it could possibly have anything to do with modelling the Sun or Moon.

By this time, Tony Freeth and Mike Edmunds had the go-ahead for their project, and Wright knew that they would be visiting the Athens National Archaeological Museum in October 2005 to image the fragments. Wright was due in Athens that same week, to present his results at a conference. It would be his last chance to claim the solution to the Antikythera mechanism for his own before the new boys muscled in on his territory. He had to get his reconstruction finished in time.

So he rushed it. With the weeks running out, the best answer he could come up with was that the lower back spiral had shown a period of four 'draconitic' months, split into 218 half days. Draconitic months follow the Moon as it crosses the plane of the Sun's orbit and are useful for predicting eclipses. He didn't know why the maker of the device would have wanted to display half-days on that dial, but this was the only astronomical period that made any sense with the gearwheels he had measured. The same result could have been calculated more easily with a fixed gear train – there was no need for epicyclic gearing – but he figured that maybe the designer's technical talents weren't matched by his mathematical skills.

Wright still couldn't see how the pin-and-slot or the 223-tooth wheel could have worked within the mechanism, so he concluded that these must have been spare parts, recycled from other devices. After all, his own model was made from old brass door plates. The exciting thing about this idea was that it would be proof that the Antikythera mechanism wasn't a one-off. Within the surviving fragments we would have the remains not just of one unique device, but two or three.

Meanwhile, Wright overlooked a broken-off shaft sitting just next to the 223-tooth turntable. He had noticed it early on and always meant to explore the idea that it might have carried a missing gearwheel. Later, Tony Freeth would place a wheel here, solving all of Wright's problems at once.

October comes and Wright arrives in Athens with his finished model, grimly triumphant as Freeth's team completes its imaging. On the day of his talk he demonstrates the workings of his device to a captivated audience. He turns the handle on the side like a magician and there's a hush as time passes before everyone's eyes, just a soft clicking sound as the Moon traces undulating circles through a miniature sky, cycling from black to silver as the golden Sun glides slowly round and the planets meander back and forth, their seemingly random paths guided by a hidden clockwork order.

Wright sees three decades of his life passing as the heavenly cycles run their course, from the young curator who was once captivated by Price's work and wished it were his own, to the man he is now, standing here with the Antikythera mechanism finally recreated and working again for the first time in 2,000 years.

His presentation was meant to be the high point of the

conference. But there is a late addition to the programme. Freeth, after finishing his imaging of the fragments and visiting the conference exhibition to glance at Wright's model, has gone home to London. But now his collaborators take the stage. In particular, Mairi Zafeiropoulou – a sturdy and rather formidable archaeologist who works with the Athens museum's bronze collection – triumphantly shows a new piece of the mechanism that she has recently found in the museum stores. It is a substantial piece of the lower back dial.

8

The New Boys

Vision is the art of seeing what is invisible to others.

— JONATHAN SWIFT

THERE WAS AN ear splitting crack, as a dazzling ten-inch spark tore across the room. The cable's free end, bursting with a quarter of a million volts, stripped a violent path through the air and discharged its devastating load of electrons into every computer in the vicinity.

'Shit,' said Roger Hadland.

It was summer 2005 and he was standing in the crowded development lab of X-Tek, the company he had founded 20 years earlier to build sophisticated X-ray imaging systems for industry. From humble beginnings the company had grown explosively, carrying him out of his garage and into smart buildings just outside the picturesque town of Tring, in Hertfordshire. The firm's success was mainly down to Hadland's passion for engineering – his speciality was thinking of new ways to build machines that could see better than anyone had seen before. He pushed his team to come up with ever smaller and more powerful X-ray sources, capable of imaging in minute detail the insides of anything from tiny electronic components to barrel-sized nuclear-waste drums.

X-Tek's precision-imaging equipment was particularly popular with microelectronics companies, who used it for quality control on their computer chips. But in 2001, just as the company had leased an extra site in order to expand still further, the dot com market crashed, taking many of X-Tek's customers with it. In the following years the company had struggled – Hadland was forced to lay off more than a third of his staff and innovation had slowed to a trickle. He planned to sell, but couldn't bear to hand over his company to callous investors who would break it apart for profit, and no one else was interested.

Then the Antikythera mechanism had come into his life, in fact taken it over, and somehow he had found himself in a daunting race against time. Inspired as never before, he was diverting company resources and turning away precious customers in an attempt to build a machine that would take the ultimate images of the ancient fragments. He wasn't sure whether to be glad or sorry – there were certainly others in the company who thought he had gone mad. On days like today, he feared they might be right.

The project had its roots back in 1998, seven years before Hadland had ever heard the fateful word 'Antikythera'. Mike Edmunds, head of astronomy at Cardiff University, was sipping tea and looking for a history project for one of his students when he came across the work of Derek de Solla Price.

Edmunds is a genial man with red cheeks and white hair, and an office with a huge, sunny window that looks out over a leafy Cardiff side street. He works on questions that go beyond the Earth, Moon and Sun, beyond anything that the ancient Greeks could have imagined: the evolution of stars, galaxies, elements and the cosmic dust that ultimately makes up planets and everything on them. On this day, however, he reined in his imagination from the edges of the universe and instead sent it 2,000 years back in time.

Reading Price's words, Edmunds was surprised that he hadn't heard about the Antikythera mechanism before, and that it didn't have a recognised place in the history of astronomy. Reviewing what was known about it would make a perfect student project.

Edmunds also described the mechanism to an old friend of his, Tony Freeth. A mathematician by training, Freeth earned a PhD as a young man exploring the strange, abstract lands of set theory. But equations weren't enough for him and he soon moved away from academia to set up a business making documentary films, working out of his book-filled home in west London. Freeth – pale and balding with a grey moustache – is more serious than the jovial Edmunds. But he's capable of quite passionate intensity when the moment seizes him. He called his company Images First, because he felt that pictures were the way to tell any story that mattered. The resulting films were earnest accounts of topics such as Alzheimer's disease, apartheid and African agriculture.

The unsung Antikythera mechanism would make a unique

subject for a documentary, urged Edmunds. Freeth, too, was amazed that it was so scarcely known, when surely it should be an icon of the ancient world. The complexity of its gearing appealed to his love of maths and logic, while he realised that making a film about the mystery might at last give him a subject that would appeal to a wider audience. He mentioned the idea to a few possible buyers, but with Price's work still standing as the last published study of the mechanism, he always got the same response: 'There's nothing new.' If he was going to sell a film on the Antikythera mechanism, he would need some fresh results.

He started to research the device, turning his attention first to Price's triumphant *Gears from the Greeks*. Like Michael Wright before him, he soon saw details that didn't add up. For Freeth, the first red flag was Price's description of the sponge divers who salvaged the Antikythera wreck: 'Only six divers were available, and because of the water depth they could not remain on the bottom for more than five minutes, which together with four minutes for ascent and descent entailed about nine minutes of submersion without air-tanks or tubes to help them.'

Nine minutes of submersion without air-tanks or tubes? Surely that was impossible. A quick search on the Internet revealed that the world record for freediving at the time was significantly shorter than that, even for lying completely still in a pool,[3] and that sponge divers at the turn of the century

3 Austria's Tom Sietas has since broken the nine-minute barrier.

would have worn full diving suits and helmets, with breathing tubes.

Freeth applied similar scrutiny to the rest of Price's paper and was soon questioning every part of his reconstruction. Like Wright, he saw through the technical bluffs and realised that Price had massaged Charalambos Karakalos's numbers to fit his theories. Freeth's fundamental concern was that Price's model involved complicated combinations of gears to calculate what were ultimately quite simple numbers. He even came up with his own 'Minikythera' design that could do exactly what Price's could, using far fewer gears. Anyone sophisticated enough to have built the mechanism wouldn't have been so wasteful.

Tony Freeth became fascinated, obsessed even, by the device. And along the way, his motivations changed. His primary aim was no longer to make a film about the Antikythera mechanism. Instead, he resolved that he would be the man to solve its long-standing mystery. The Antikythera bug had infected its next victim.

He knew that his best chance – of cracking the mechanism and making a compelling film – lay in getting the best possible images of the gearing. Price's radiographs were lost. Freeth had heard from Mike Edmunds that some chap from the Science Museum had also X-rayed the fragments, but that was a decade ago, and as far as Freeth could tell nothing had ever come of it. Probably nothing ever would, he thought dismissively. From what he knew of the crude tomography technique the man had used, his images would be fiendishly difficult to interpret. Freeth wanted something altogether more sophisticated.

He scoured the scientific literature for reports of the latest imaging technology, and found two articles that promised a way forward. First, he saw on *Nature*'s front cover a colourful picture of a goldfish's innards, showing its feathery ribs and even its internal organs in subtle yet sharp detail. Normally X-rays are blocked by bone but pass straight through a body's soft tissue, which is why a conventional X-ray image of a person shows just their skeleton. This paper's Australian authors had perfected a technique that detects the change in the phase of X-rays (how the highs and lows of the different waves line up) as they pass through different types of material, achieving intricate images of objects that wouldn't normally show up at all.

Freeth emailed the researchers to ask whether the method would work on the Antikythera mechanism. But the reply came back explaining that unless they were chopped into thin slices, the dense bronze of the fragments would likely block the X-rays altogether. Instead the authors suggested that Freeth investigate a technique called microfocus X-ray imaging, which uses a powerful but tiny X-ray source so that the radiation it emits can be focused very precisely to give a high-resolution image. The latest machines used digital detectors rather than film, which recorded the precise amount of radiation hitting each pixel and fed it directly into a computer.

Microfocus X-ray imaging sounded like the perfect way

to gain clear images of the tiny inscriptions and gear teeth of the Antikythera fragments. So Freeth found his way to Hadland's X-Tek. The company's X-ray sources were only a few thousandths of a millimetre across, yet capable of producing strong beams of radiation. Freeth spoke to the sales manager, who agreed to help almost immediately. The project sounded straightforward – well within the capabilities of X-Tek's equipment – and the press coverage the results were likely to get would be a great marketing opportunity for the company.

The second article, in *New Scientist* magazine, described a technique that was being used to decipher some of the oldest writing in the world. These were cuneiform inscriptions; the wedge-shaped letters that the people of Mesopotamia – today's Iran, Syria and Iraq – pressed into wet clay as far back as the fourth millennium BC. The surviving tablets, though amazingly well-preserved for their age thanks to the dry sand that has cradled them for so long, are worn and faded, with many of the inscriptions now quite illegible. The article described how a young researcher working at Hewlett-Packard's labs in Palo Alto could transform photographs of these dull tablets into sharp, glossy computer images with inscriptions that practically jumped off the screen.

The researcher was a floppy-haired Californian called Tom Malzbender, and decoding invisible messages from past civilisations was the last thing that he had expected to find himself doing. He had actually been trying to develop more realistic computer graphics. It was easy enough to create a computer image of an object, a knight in shining armour, say, and have it clank around the screen. You just wrote a piece of code to model the texture of the material you wanted, then wrapped it round the appropriate geometric shape for each frame of the animation. But for a more realistic effect, Malzbender wanted to model how different materials look as they move past a light source. The reflections on that suit of armour will change as the knight walks under a chandelier or steps out into the sunlight. And the changing appearances of materials that reflect light in more complicated ways – curly hair, for example, or a crumpled newspaper – are even harder to calculate.

Tom Malzbender's relaxed exterior hides an acute technical expertise, combined with the sort of mind that refuses to see the world according to normal rules. Instead of trying to come up with more sophisticated mathematical models, as his competitors were doing, he decided to let nature do some of the work for him. He realised that if he took lots of digital photographs of the same object lit from slightly different directions, he could feed those images into a computer and measure how each pixel in the picture varied – creating a virtual map of the object's response to light. So he built a light-proof dome, with a camera fixed inside at the top, pointing downwards. The rest of the dome's inner surface was covered with 50 flashbulbs, wired to fire one after the other, each time the camera shutter flicked open. The equipment looked like a home-made flying saucer, but it worked just as he hoped. Once the computer had done its analysis it was like having a magic wand that imposed his will on to the screen – he could dim the lights, turn them up or wave a spotlight over the object in the image with the twitch of a mouse. He could even create effects that would be impossible in the real world, such as lighting a surface from behind, or hanging a lamp inside the thinnest scratch.

In 1999 Malzbender attended a lecture given by an archaeologist who was trying to decipher ancient cuneiform tablets. Realising that his imaging technique might help to enhance the faded inscriptions, Malzbender offered his services and was given a crumbling tablet on which to test out his idea.

After photographing it, he played around with the resulting image. As well as changing the position and brightness of the light source, he realised he could ask the computer to change the way that the object itself reflected light. For example, making every pixel reflect light more strongly worked like coating the dull, dusty surface in glossy metal, making every defect as obvious as a scratch on a brand-new car. And by setting each pixel to only reflect when the light was pointing directly at it, he could make tiny marks jump out like gleaming stars against the dead black of night.

The results were stunning. The tablet was a draft contract, written around 3100 BC for a Sumerian slave trader called Ur Ningal. In Malzbender's images the inscriptions on the worn, crumbly surface appeared as clear as shining crystal, including some that it hadn't been possible to read at all before. He even saw the fingerprints of the scribe who held the clay while it was still wet.

As soon as Tony Freeth saw the pictures he wanted the dome for his Antikythera project. He e-mailed Malzbender in 2001 to explain his idea, but the graphics expert didn't think much about it – he received a lot of requests to use his dome to image various objects, and most of them never came to fruition. The next year, however, Malzbender happened to spend three months on sabbatical in Bristol and while he was in Britain he travelled to London to meet Freeth for lunch in the refined surroundings of the National Gallery.

Malzbender was keen to work out whether Freeth was serious about the proposed project. The man seemed quiet, almost awkward – very British, Malzbender thought. But five minutes into their conversation, he was impressed: he was used to explaining the technicalities of his algorithms over and over to duller minds than his own, but Freeth understood every detail instantly. Ten minutes in, there was nothing more to say. Malzbender knew that he would be going to Athens.

Freeth now had two companies willing to help him get more information out of the Antikythera fragments than ever before. X-Tek's state-of-the-art technology promised to make sharp X-ray images of the gearwheels inside the device. And Hewlett-Packard's revolutionary light mapping would help him to read previously hidden surface inscriptions.

But there were two big problems. Freeth didn't have any

funding for the project; although both companies had promised to give their time for free, he still needed to raise enough money to ship all of the necessary people and equipment to Athens. Worse, he didn't have permission to study the fragments.

Officials at the Athens National Archaeological Museum had turned down his request. A campaign was hotting up to persuade Britain to return the Elgin marbles to Athens in time for the 2004 Olympics there, so a British researcher wanting access to a Greek artefact was not in a good bargaining position. But Freeth would not take no for an answer. He calculated that if he could get some prominent Greek scientists on board the museum might change its mind, so he set about putting together a collaboration that would be influential enough to push his plans through.

Trying to solve the mystery of the Antikythera mechanism gave Tony Freeth a sense of purpose that neither his maths nor his films had ever given him, and over the next couple of years all of his other projects fell away as he devoted more and more time to his quest. He organised petitions, published papers and wrote grant applications. He set up an e-mail discussion group devoted to gaining access to the fragments, and lobbied everyone he could think of for support. And he never stopped going over Price's old work.

Gradually a team came together. The first recruit was John Seiradakis at the University of Thessaloniki, one of Greece's most eminent astronomers; then came Seiradakis's friend Xenophon Moussas, an astrophysicist at the University of Athens with close links with the National Museum; and there was Agamemnon Tselikas, director of the Centre for History and Palaeography in Athens and an expert in reading ancient texts. Finally, Mike Edmunds provided the necessary academic credentials from Britain. They were a band of brothers, Freeth thought proudly. He would lead them to victory.

John Seiradakis and Mike Edmunds, as the most senior scientists on the team, applied to the National Museum with the full force of their joint academic reputations. Edmunds also applied for grant money from the Leverhulme Trust, an organisation set up by William Lever (founder of the company that eventually became Unilever) to fund unique and interdisciplinary projects that might not have a chance of support elsewhere. The team finally won its money early in 2005.

Just two weeks later, the answer came back from officials at the National Museum. Despite the presence of the Greek collaborators, the answer was still no. As far as the museum was concerned, the Antikythera fragments had already been X-rayed – twice – and work on the latest round of data was still in progress. There was no reason to put the crumbling pieces at further risk by imaging them again.

Freeth refused to consider defeat, so he changed his plan of attack. The only organisation with the power to override the National Museum's decision was the Greek Ministry of Culture. Xenophon Moussas took over the fight.

Moussas is a gentle, soft-spoken man, not normally the type to make trouble. But he is fiercely proud of his Greek heritage. When he was a child growing up in Athens he used to love going to the National Museum – just as a young Michael Wright had been enchanted by London's Science Museum. Moussas never tired of telling Freeth and the others how he used to stand in front of the Antikythera fragments in their glass case, marvelling at the sophistication of the ancient mechanism and losing himself in daydreams about his forefathers, the ancient Greeks. Away from the museum, he'd look up at the night sky and imagine those astronomers from long ago who were equally inspired by the same sight.

As Moussas grew up, he never lost his fascination with the skies. He became a physicist, specialising in studies of the Sun. Many of the projects he worked on recalled the Greeks' past glories, from NASA's *Ulysses* spacecraft on its distant wanderings over the Sun's poles, to *Artemis 4*, a radio telescope positioned on the plain at Thermopylae, where King Leonidas and his small band of Spartans had once fought the full force of the Persian army.

Moussas took his new role on the Antikythera team seriously. If only he could speak to the Culture Minister he could explain the merits of the project – the quality of the equipment and the eminence of the researchers who were on board, and the importance of understanding the mechanism for Greece! Only the Antikythera mechanism could show the true extent of the ancient Greeks' achievements

– in science and technology, not just art and battle. If the minister would only hear the story, he would surely force the museum to give them access.

Moussas phoned the Ministry of Culture again and again, but he could never get anyone to put him through. Perhaps his persistence eventually wore the secretaries down, however, because after 40, 50, maybe even 60 calls, he was finally granted a meeting with the Deputy Culture Minister Petros Tatoulis and his wife Sofia.

Moussas's words fell on fertile ground. As well as a passion for archaeology (since his appointment in 2004 Tatoulis had been leading efforts to pressure the British Government into returning the Elgin marbles), it turned out that the couple had a keen amateur interest in ancient astronomy. Not to the entire delight of the National Museum staff, Tatoulis granted permission for the team to study the fragments in September 2005. Everything was in place. With just a few months to wait, an elated Tony Freeth told his collaborators at X-Tek and Hewlett Packard to prepare for the trip to Athens.

It wasn't to be quite that simple. X-Tek's technology had moved on since Freeth originally approached the company in 2001. Roger Hadland's engineers had developed equipment capable of combining microfocus X-ray imaging with a technique called computed tomography, or CT. This is basically a much more sophisticated version of the method Michael Wright had used with his home-made tomography cradle. Rather than taking a series of flat radiographs, CT can produce a three-dimensional reconstruction of the entire object, so with the right computer software you can fly right through its insides, exploring every hidden corner. The technique uses an X-ray beam that spreads out from the tiny source in a cone shape, before passing through the target object and then hitting a square detector. Each pixel of the detector measures the precise amount of radiation that hits it. From this, you can draw a series of straight lines through the object, from the X-ray source to each individual pixel, and work out exactly how much radiation was absorbed by the object along each line. That doesn't tell you much by itself. But then you rotate the object a tiny bit and do the same thing again, in fact you do it thousands of times until you have an image from every possible angle.

Depending on the position of different structures inside the object, each image taken as the object is turned will cause a slightly different pattern of radiation to hit the detector. And that gives the computer enough information to work out the exact arrangement of the object's insides – like filling in the squares in a game of X-ray sudoku.

Freeth had realised that three-dimensional CT would be the ultimate tool for understanding the internal details of the mechanism, and when he asked the sales team if X-Tek could use it to image the Antikythera fragments, they had said yes. But with the project now actually about to happen, Roger Hadland himself finally got to hear of the promises his team had made.

He was immediately concerned. The Antikythera fragments were made of bronze, a very dense metal, and the fragments were large by the standards of microfocus CT – the largest was more than 18 centimetres across. With normal X-ray images it doesn't matter if some parts of the target object block X-rays completely, they just come out black in your final image. But with CT, the team would need information from every pixel, even when probing the fragments end on, or the computer wouldn't be able to accurately reconstruct the internal details.

X-Tek's microfocus sources work on the same principle as larger ones – an electron beam hitting a tungsten plate knocks electrons in the tungsten atoms up to higher energy levels, which then relax and spit out high-energy photons – X-rays – in the process. The power of an X-ray source is given in volts, which is basically a measure of the strength of the electron beam producing the X-rays. X-Tek's smallest sources weighed in at 225 thousand volts (225 kV).

No one in the world could better that, but Hadland estimated that to guarantee sharp CT images of the Antikythera fragments would take around 450 kV. Such a machine simply didn't exist.

Roger Hadland didn't want his company associated with anything less than perfect data. 'I'll ring Tony and tell him we can't do it,' he told the sales team. He had never spoken to Freeth, but he expected it to be a brief call – Freeth would express polite disappointment, Hadland would express regret, and X-Tek's foray into ancient history would be over.

'*You have to do it!*'

It was not a request from Freeth, but a forceful statement of fact. There followed a torrent of admonition, explanation, persuasion and drama the like of which Hadland had never experienced. One hour later, he had made a 180-degree turn. He was now convinced that imaging the Antikythera mechanism would be a once-in-a-lifetime opportunity to discover the secrets of one of the most important artefacts that survives from the ancient world. He was also starting to think about what a tremendous opportunity the project might be for X-Tek.

Imaging the mechanism would require a completely new machine, with X-ray beams twice as powerful as any the company had developed so far. Hadland had long wanted to design such a very high-voltage X-ray source, because it would allow the company to expand into a badly needed new area of business – imaging the turbine blades in aircraft engines.

In a turbine engine, air is heated to searing temperatures by burning fuel, then as it expands it rushes past the turbine blades, turning them and driving the engine. The gases are hotter than the melting temperature of the blades, so they have cooling channels inside to dissipate the heat. If there is any defect in these channels, the blades can fail, with devastating results for the aircraft. If X-Tek could build a microfocus X-ray machine capable of 450 kV, it would be powerful enough to image the tiniest cracks in objects as big as turbine blades.

The problem was, such an ambitious design project typically takes two or three years. Since the bottom dropped out of the microelectronics market, Hadland just couldn't afford to invest effort in a project for that long without any immediate financial returns. But the more he thought about the Antikythera project, the more his heart ruled his head, and the excitement of invention filled him, as it had when he first built up his company. Freeth had been given a specific time slot at the National Archaeological Museum: September 2005. It was already June, giving Hadland less than four months. It was a ridiculously short space of time, but what if he blitzed it? He could put everything the company could spare into the project for a few months, image the fragments, then dive into new markets with his world-beating X-ray system before the year was out.

Hadland assigned almost all of his R&D staff to work on the new machine. They needed to double the voltage of X-Tek's existing microfocus X-ray source, so they decided to place two of these sources back to back, making one positive and one negative, so that the resulting electron beam would be accelerated into the tungsten target with twice the energy.

It sounded simple, but there was a lot of work to be done. The company already had a negatively charged X-ray source that would do the job, but positively charged ones are much more difficult to make. Seven engineers, including Hadland himself, would have to develop a new electron gun, a new high-voltage generator, and a new controller that could operate each of the generators both individually and in tandem. Another three people would work to update the computer software needed to reconstruct three-dimensional images from the collected data.

With September ticking ever closer, Tony Freeth waited impatiently. Since his team had made public its plans to study the fragments, Michael Wright had started publishing his work. Slowly but surely, Wright was solving one part of the mechanism's structure after another. Freeth was convinced that Wright had been doing everything he could to block his team's access to the fragments. And he was seriously worried that by the time he got to throw all of his technology at the problem, there wouldn't be anything left to discover.

He did have one ace up his sleeve, however. A member of staff at the museum, Mairi Zafeiropoulou, had been gathering together the Antikythera fragments for the team to image them. The holdings of the National Museum had still never been fully catalogued, so the precise contents of its storerooms were always something of a mystery, even to the staff. The largest Antikythera fragments – A, B and C – were easy enough to find, because they were on display. But Zafeiropoulou had no idea where Fragment D was. After much searching, armed with Price's photos, she eventually found it in an unmarked wooden case. She also located lots of tiny scraps, which seemed to have been scraped off the larger fragments as they were cleaned, which might contain vital letters from the mechanism's inscription.

She also found Fragment E, which had been discovered

in the stores by the curator Petros Kalligas in 1976. And then, wonderfully, lying in a box alongside an old piece of wood, she found a green, corroded lump that didn't match any of the old photographs. It was a few centimetres across and completely untreated, covered in limestone with worm tracks still winding over its surface. Underneath it was just possible to make out portions of concentric circles – matching those on the back dials of the Antikythera mechanism. She called the new piece Fragment F. imaging it would provide essential clues that no previous researcher had seen.

September arrived, and X-Tek's machine still lay in pieces on the workshop floor. But Tom Malzbender, along with colleagues Dan Gelb and Bill Ambrisco, flew across the sea from California with their flashbulb dome packed in a crate.

Once at the National Museum, they were taken down to a basement to do the imaging. It was a bit of a shock after the well-equipped labs of Hewlett-Packard – the room was basic and bare, with antiquated wiring running exposed around the wall, and the lack of air-conditioning made it stifling. Malzbender was the only one allowed in the room with the Antikythera fragments (watched over by a museum official), so he rolled down the shutters and turned off the lights, while Tony Freeth and the others waited eagerly next door. Malzbender felt privileged to be alone with the ancient mechanism, but for once he was just a little nervous. He had travelled a long way to make this happen. It was no time for his equipment to go wrong.

Over the next five days Malzbender took more than 4,000 images of 82 fragments; so many, in fact, that it nearly wore out his camera. His dome hung vertically, so that when each piece was in place, he had to stoop to line up the shot before pressing the shutter to release 50 flashes. Then he downloaded the images on to a memory stick and walked to the next room where Dan Gelb had his laptop and the software needed to reconstruct the images. At first, the composite pictures appeared on the computer screen as high-resolution colour photos; a series of sharp splotches of dull green and beige.

Tony Freeth hovered in the background, leaning over Gelb's shoulder as he coaxed the software to work its magic. With a few keyboard strokes, Gelb took the fragments out of this world and into one not bound by the laws of physics, transforming them into beautiful drops of mercury that hung in the black background of space and glinted silver in the distant Sun. These mercury asteroids were covered in bold, clear, impossible writing: a message through time from one civilisation to another.

At the end of the week Malzbender and his colleagues packed up to go home, but Roger Hadland and his team were still working late nights in Tring. Fortunately, Freeth had managed to persuade the reluctant museum staff to give them another slot with the fragments in October. So the X-Tek team burrowed themselves away in the development lab, while customers' requests fell on deaf ears and one order after another was lost. If the Antikythera project didn't work,

Hadland thought, he wouldn't have much of a company to go back to.

With just a week left until the final deadline, he was starting to panic. The high-voltage generator wasn't working. It was meant to conjure a voltage of 225 kV – nearly a quarter of million volts – but the circuit was only registering a measly tenth of that. Eventually, in desperation, the engineers unhooked the cable from its X-ray source and switched it on to inspect the generator's workings more closely. They weren't concerned about the disconnected cable – 20-odd thousand volts is low for equipment like this, and should be quite safe in the open air.

Famous last words. There followed an almighty crack, like a pistol shot, as a huge spark, almost a foot long, leapt from the cable's end. Then a shocked silence as each man present briefly considered his own mortality – if it had hit someone, such a huge electrical discharge could have killed them. But after a few seconds, Hadland started to grin. That proved the generator was working fine. Someone must have put the wrong resistor in the circuit, causing the voltage to read ten times too low.

Unfortunately, the explosion had just destroyed the computers they were using to run the machine. Then the second deadline came, as did the lorry booked to carry the machine to Athens. The X-Tek team worked on and the Greek lorry driver had to wait in Tring for two days, sleeping in his cab while they finished rebuilding the system.

Late in the evening on the second day, they were finished.

The huge lead-lined cabinet contained an X-ray source that was smaller and more powerful than any other in the world. Hadland called it BladeRunner, after the turbine blades he hoped it would one day be used to test.

Pride soon gave way to exhaustion. It took a couple of hours to pack up the machine, literally building a crate around the cabinet with timber and plywood, then they had to load it on to the lorry in the early hours of the morning. The cabinet alone weighed nearly nine tons, and the fork-lift truck they had hired couldn't take the weight without tipping over. Putting another ton of lead on the back of the forklift solved that problem, but then it didn't have enough power to move forwards, so they had to use a second fork-lift to push it along. The various X-ray sources and detectors added another couple of tons to the lorry's load, and finally it set off for Athens.

It would take the driver five days to get there, heading slowly south through Italy to the port of Brindisi, before rolling on to a ferry bound for Greece. Meanwhile Roger Hadland and Tony Freeth jumped on a plane, meeting the driver just outside Athens. The lorry was 20 metres long – perhaps not quite as big as the Antikythera ship, but still, it became apparent that it wasn't going to fit through the narrow streets of the city. Its precious cargo would have to be transferred to a smaller truck. It took them all day to find a company with forklifts that were up to the job, followed by a second late-night loading exercise.

The next day a police escort ensured that they reached

the National Archaeological Musuem without further mishap. The trees outside the pretty, white building left dancing shadows in the early morning sun, but Hadland's attention was focused solely on the workers he was directing; he felt like a nervous parent as the plastic-wrapped crate was inched down from its truck and through the museum's black iron gate. By the time it reached a gentle ramp into the building itself, the sun was high overhead. It took the rest of the day and three forklifts chained together – two pulling and one pushing – to shift the enormously heavy box up the slope. Hadland wondered if it had been this much work building the pyramids.

The rest of the week was spent getting everything wired up, taking some conventional X-ray images, and calibrating the system. Everything was still at the prototype stage, so a mass of scrappy wiring connected the equipment in the cabinet to a couple of industrial PCs, sat on packing cases for tables. The rest of the room was filled with spare generators, X-ray sources, repair kits, cabling – everything Hadland could think of to bring, just in case any of the components failed.

One of the first pieces of the Antikythera mechanism to be examined was the almost-lost Fragment D with its single cogwheel. It should provide a simple test of the CT. They started up the equipment and a cone of invisible X-rays splayed out from the source, through the fragment and on to the detector behind. Rather than risk damaging the fragment by stopping and starting for each image, it was fixed to a slowly rotating turntable. It inched around once in an hour like a clock's minute hand, while the computer recorded ten images for every degree – more than 3,000 over the entire 360-degree rotation.

When the circle was complete, it took about an hour for the computer to assemble all of the data into a three-dimensional volume. Finally, X-Tek software engineer Andrew Ramsey pulled up the images on his computer screen, with a crowd of people including Freeth, Hadland, Moussas and various curious museum staff jostling behind him. One face missing was Mike Edmunds, who had stayed in Cardiff.

There was silence. The surface images from Malzbender's team had been stunning, but everyone knew that for the project to be a success they needed to see inside; they needed to see the internal workings. Ramsey scrolled down through the depth of the fragment. At first all they could see was a blur, but then a crackling sharp gearwheel emerged from the fuzz, as if being hauled up out of grey sand. It was better than any of them had dared hope. The letters 'ME' had been scratched into the side of the wheel. It was like a signal from the past, an 'I WOZ ERE' from 2,000 years ago. Suddenly, they felt a direct, almost physical connection with this ancient machine, and with whoever had carved those letters so long ago.

Then Freeth started to laugh. 'Somebody e-mail Mike and tell him we've found a gearwheel with his initials on!'

After that they worked through the fragments one by one, working up to the biggest Fragment A, with its characteristic four-spoked wheel, learning how to get the best results out of the machine as they went, and all the time getting more and more excited by the details they found. It became clear that there was hardly any intact bronze left inside the fragments – something that Michael Wright could have told them if they had asked, after his experience 15 years earlier when the delicate zodiac scale had snapped in his hands. But at that moment Wright was alone in his workshop preparing for his own trip to Athens, desperately crafting the finishing touches on his model of the mechanism.

The state of the fragments meant that there was less dense metallic material inside than the team had thought, making them much more transparent to radiation. That meant the pictures were even sharper than the researchers had hoped, with resolution in some places down to just a few thousandths of a millimetre. Every surviving part of the mechanism – every tooth, peg, shaft and pin – could be seen in pure, crisp, breathtaking detail.

The team worked as long as the museum staff would let them each day (although that was still nothing to how hard Michael Wright had laboured in that cramped darkroom, Eleni Magkou thought, as she watched the team's colourful comings and goings). Each evening the group would meet in the lobby of their hotel and they'd go for dinner in jubilant mood, raising their glasses to the cry of 'More gears!'

As an engineer, Roger Hadland was especially dumbstruck by how familiar the components inside the mechanism were. It was like opening a textbook that displayed many of the techniques and arrangements of gearwheels that he saw around him every day, in automatic devices from car windows to camera shutters. The other surprise was that the CT revealed lots of hidden inscriptions, which had been invisible even to Malzbender's team, because they were completely covered by limestone or the products of corrosion. The hollows of the engraved letters were less dense than the corroded metal around them, so as Andrew Ramsey scrolled just beneath the surface, even the tiniest inscriptions showed up beautifully.

Progress was steady, although they had trouble finding enough computer hard drives to store all of the data. Each scan incorporated 3,000 images of 12 megabytes each, adding up to 36 gigabytes per scan. Overall the team produced nearly a terabyte of data. That's enough to fill a library, and to print it all out on paper would require pulping a small forest of trees.

Time was also in short supply. The National Museum had given the team a strict slot of three weeks to finish their imaging and get out of the building. As the finish date drew closer, they had their data on the larger fragments, but realised that there wouldn't be enough time to image all of the smaller scraps. These might not even all come from the mechanism, but without putting them in the scanner it was impossible to know which ones contained a crucial clue. Xenophon Moussas disappeared off to a little art shop down the road and came back with several polystyrene cylinders, out of which he gouged cradles that fitted the pieces snugly.

If they fixed the cylinder to the turntable, they could scan several fragments at once. It was clumsy, but it worked; they got the images they needed just before they were thrown out. As a determined Michael Wright completed his Athens lecture and carried his precious model home to London, BladeRunner began its slow lorry ride back to Tring.

9

A Stunning Idea

It's an absolutely unbelievably stunning and sophisticated idea. I don't know how they thought of it. We're just following in the tracks of the ancient Greeks.

— TONY FREETH

THE CENTRE FOR History and Palaeography is on the first floor of a grey, stone office building, squeezed into an Athens back street just a couple of minutes walk from the old Roman marketplace and the Tower of the Winds. It doesn't look like much from the outside, but within its doors the shelves overflow with piles of old books and manuscripts, and humidifiers fill the air with moisture, so that the delicate paper doesn't shrink or curl.

In the corner a dried goatskin is stretched taut over a wooden frame – a demonstration for visitors of how parchment is made – and on one wall a series of posters explains how the physicist Yanis Bitsakis decodes ancient palimpsests. These are old texts on parchment that has been washed and scraped clean, so that a new text can be written on top – a common practice in the Middle Ages when parchment was a scarce raw material. Sometimes these scraped-away traces are all that's left of important ancient documents. A tenth-century copy of several of Archimedes' mathematical writings, for example, was recently found beneath the words of a medieval prayerbook.

Yanis Bitsakis uses his technical expertise to uncover stories where others wouldn't even bother to look. To reveal the hidden layers of ink on a page, for example, he photographs it with a camera that detects different colours – or wavelengths – of light. Different inks reflect characteristic combinations of wavelengths, meaning that when he displays the images on a computer screen he can turn the different colours up and down – even those not normally visible, such as ultraviolet and infrared – looking for the perfect mix to muffle the overlaid writing and bring out the lost messages beneath.

Now he has a new subject. Tony Freeth's team has hired Bitsakis to work through the thousands of computer images of the Antikythera mechanism from Hadland's BladeRunner machine and Malzbender's flashbulb dome. With his long dark hair tucked behind his ears, the physicist checks each separate slice of the CT data, and painstakingly tries out all of the lighting conditions he can think of on Malzbender's photos, looking for every last trace of visible lettering. Then he passes the most useful images on to the philologist Agamemnon Tselikas, the director of the centre, who does his best to read them.

Agamemnon Tselikas (Memos to his friends) is a large

but gentle bear of a man, excellent company and fond of life's pleasures. But for three months, he spends his evenings alone with images of the Antikythera mechanism, working in silence every night from around eleven until the early hours of the morning. The letters are tiny, some less than two millimetres high, and all squashed together without spaces with no clue as to where each word starts and finishes. He drinks thick, black Greek coffee as he scrolls from one slide to another, trying to get inside the head of the machine's maker, so that he can decipher his words.

The results start to flow almost immediately. The very first inscription that the two men read – on the back of the mechanism quite near the top – is 'ΕΛΙΚΙ', which means 'spiral'. It appears within the phrase: '. . . the spiral divided into 235 sections . . .' In London, Tony Freeth nearly jumps out of his armchair when Bitsakis phones to tell him. The snatch of text – clearer than any that Derek de Solla Price was able to translate – shows that the inscriptions do contain operating instructions. And it confirms what Michael Wright has been saying about his measurements of the upper back dial. Here is independent evidence that the dial was a continuous spiral rather than a series of concentric circles, and that it was divided into the 235 synodic months of the 19-year lunisolar cycle. Whereas the front dial gave the day of the year, this back dial allowed the user to track months and years over much longer periods of time.

Freeth is also poring over the inscriptions. On the front dial, he reads 'Parthenos' (Virgo) and 'Chelai' (Libra), as

Price did before them. But diving under the surface with the CT he sees the next sign 'Scorpio', further proof of the zodiac scale running clockwise around the dial. He is also able to add a few letters here and there to the parapegma text on the front of the device, and see more reference letters on the dial, enough to show that the calendar probably ran through the alphabet twice.

One of the most extensive, newly revealed texts comes from the front door plate. The researchers only have the middle part of each line, but from the surviving words they can tell that it is clearly discussing the planets: 'Venus' is mentioned, and 'Mercury', plus the 'stationary points' that Price saw, and there are also some numbers that might relate to the distances between the planets and the Sun.

And on the back of the mechanism they find a long list of operating instructions. Mechanical terms such as 'trunnions', 'gnomon' and 'perforations' are mixed with astronomical references. As well as the spiral inscription, the numbers 19 and 76 appear – the number of years in a Metonic and Callippic cycle, respectively. And there are references to 'golden little sphere' and 'little sphere', probably referring to the Sun and Moon pointers on the front zodiac display.

The text near the lower back dial includes the number 223, possibly the word 'Hispania' – the earliest known mention of Spain as a country – and a number of other geographical references (some new, some read earlier by Price), such as 'from the South', 'towards the East' and 'West-North-West'. The CT also shows that the little dial next to the lower back spiral is divided into three. One of the three divisions seems blank, while the other two are mysteriously inscribed with letters that stand for the numbers 8 and 16.

Finally, Tselikas spends a long time staring at the wheel of Fragment D, with its letters 'ME'. They are tantalisingly clear, but he can't say for sure what they mean. The inscription might be short for the Greek word 'messon', which means median or 'the one in the middle'. If the letters were meant to be read as digits they might stand for the number 45. Or perhaps they are the initials of the maker himself.

In all, Bitsakis and Tselikas have more than doubled the number of legible characters on the mechanism to well over 2,000, out of what may originally have been around 20,000. Dating the text precisely is difficult, because the tiny carved letters are quite different to any other writing that survives from the period. But the style is in line with around 100 BC or perhaps a few decades either way. What Tselikas became sure of during his lonely nights, however, is that the Antikythera mechanism was not meant to be used by the person who built it. Everything about its workings was explained step by step. Rather than being an astronomers' instrument or workshop tool he feels it had to be a luxury item made for a wealthy, non-specialist owner.

While the inscriptions were being translated, Tony Freeth turned his mathematician's mind to the arrangement of the gearwheels. He had been following every detail of Michael

Wright's publications and knew Wright believed that the gearing on the front of the device, now lost, had modelled the varying motions of the planets, as well as wobbles in the speed of the Sun and the Moon. The inscriptions found by Bitsakis and Tselikas supported the idea that the planets had been shown, but without the lost pieces Freeth felt it was impossible to prove the case either way. He decided to restrict himself to the parts of the mechanism that had survived.

The first task was counting the gear teeth. Instead of counting by eye, Freeth and Edmunds then used a computer programme to crunch the maths for them, making the tooth counts more certain than ever before. He soon confirmed Wright's reading of the gear train that drove the Sun and Moon pointers, the Moon-phase indicator on the front dial, and the train leading to the upper back dial to show the 19-year cycle. But he also discovered a striking new feature that explained how the spiral reading was displayed. Wright had suggested that marker beads might have been moved around the spirals to mark specific dates. But with the CT images Freeth could see the details of the end of the surviving pointer: it had an extendable arm with a pin at the end, which had travelled around the spiral groove like the stylus on a record player. It would have taken 19 years to travel all five turns of the spiral, then the arm could have been lifted and set back at the beginning.

Then he came to the lower back dial. Wright's interpretation – that the spiral showed draconitic months divided into 218 half days – didn't feel right. But here Freeth had a distinct advantage, because Fragment F – found by Mairi Zafeiropoulou in the museum stores – was a key section of this dial. It came from the bottom right corner of the mechanism, and it showed parts of all the rings of the spiral, with their scale divisions. This allowed him to make a much more confident count of the divisions than had been possible before. It came to 223.

Etched into the segments on this scale, Freeth identified 16 blocks of characters or 'glyphs' at intervals of one, five, and six months. Some of the glyphs contained the character 'Σ', some contained an 'H', and some had both. Then there was what looked like an anchor sign followed by a number, and one more letter at the bottom. A few of them were visible and had been seen by Price as well, but all the others were hidden under the surface and could only be read with the CT.

All the evidence pointed in one direction – that this dial was used for eclipse prediction. The geographical references and directions inscribed around the dial fitted that idea; solar eclipses occur only at certain locations, and ancient eclipse observations often mentioned the direction from which the shadow approached. Lunar and solar eclipses tend to occur at intervals of one, five and six months from a particular start date. And the pattern repeats almost exactly after what's called a Saros cycle: 223 months. This is what the 223 divisions of the spiral had to be showing.

The cycle works because for an eclipse to happen you

need three things. First, the Moon has to be either at full Moon (for a lunar eclipse, where the Earth passes between the Sun and the Moon) or new Moon (for a solar eclipse, where the Moon passes between the Earth and the Sun). So if an eclipse happens, you can only get another identical one after a whole number of synodic months has passed. Second, the Moon's path has to be crossing the plane of the Earth's orbit around the Sun, so that all three bodies are in a line. If the Moon orbited in the same plane as the Sun, we'd get eclipses every new and full Moon. But it's actually tilted with respect to the Sun's orbit by about five degrees, so you only get an eclipse if the new or full Moon happens just as the Moon crosses the line of the Sun. The time period between these crossings is the draconitic month.

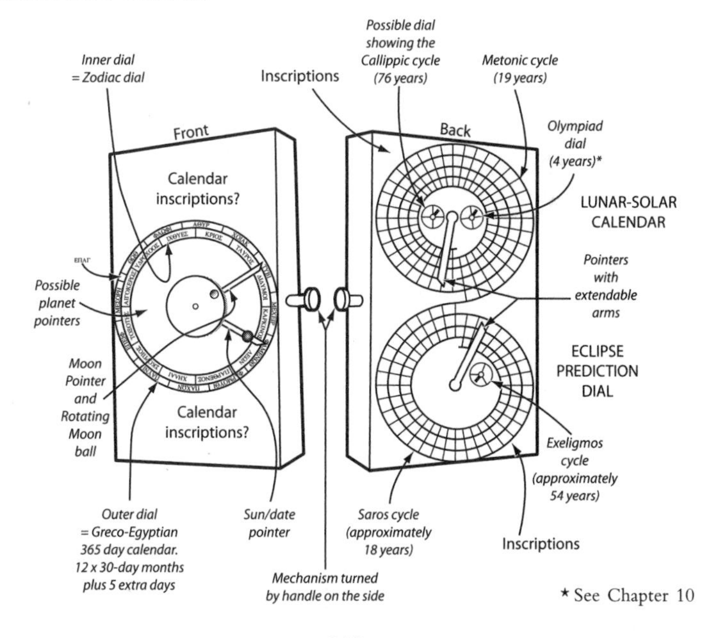

* See Chapter 10

After a period of time in which both a whole number of synodic months and a whole number of draconitic months has passed, eclipses will start repeating themselves. The Saros cycle is such a period, equal to almost exactly 223 synodic months, and 242 draconitic months.

There's one other reason why the Saros cycle is particularly good for predicting eclipses. Because the Moon travels around the Earth not in a perfectly circular orbit but in an ellipse, its size and speed with respect to us vary as it goes around. The Moon looks bigger and faster when it's at the nearest point of its ellipse, and smaller and slower the further away it gets. The time taken for one complete ellipse is about 27.5 days and this is called an anomalistic month. It's very slightly longer than a sidereal month in which the moon returns to its same place with respect to the background stars, because the direction of the furthest point of the ellipse is gradually moving around the Earth, at a rate of about once every nine years. Where the Moon is in its ellipse determines how long an eclipse lasts (if the Moon is going faster it'll be over quicker) and whether we see a total eclipse or just an annular one (if the Moon is far away, it isn't big enough to blot out the whole Sun). A Saros period contains almost exactly 239 anomalistic months. So every 223 synodic months – or just over 18 years – eclipses don't just happen at the same time; the characteristics of each eclipse will be similar as well.

The Babylonians first came up with the Saros period, although that isn't what they called it. As with the 19-year

Sun-Moon cycle, they didn't understand the theory behind why the pattern worked, instead they learned it from centuries of observations, all carefully noted down on clay tablets like the ones Tom Malzbender uses his flashbulb dome to read. Lunar eclipses in particular were one of the most powerful omens that could occur. They generally meant that some terrible event was about to happen, such as the death of a ruler, which could only be averted with the appropriate ritual or sacrifice. Knowing in advance when an eclipse was going to happen helped preparations no end, and made sure that the priests didn't miss one, even on a cloudy day.

Around the second century BC Greek astronomers found out about the Babylonians' eclipse cycle. Eclipses didn't have the same superstitious meaning for the Greeks as they did for the Babylonians, but Greek astronomers had other reasons to be interested in their timing. Lunar eclipses give a regular and precise record of when the Moon is exactly opposite the Sun in the sky, so they used the data to derive numbers for their geometric models of the Moon and the Sun.

There is one problem with the Saros period, however, which is that it doesn't add up to a whole number of days – instead it lasts 6,585 and a third days. This means that during any particular Saros cycle, eclipses occur eight hours later than in the one before. And solar eclipses, which are only visible from certain locations, occur 120 degrees further west, because the globe has had time to go through an extra third of a revolution. So the Greeks came up with the idea of a longer cycle consisting of three Saros periods or 54 years. They called it the Exeligmos cycle (from the Greek for 'revolution'). This does add up to a whole number of days, so after an Exeligmos period the eclipses repeat in almost exactly the same pattern.

This explained why the subsidiary dial was divided into thirds. After each 18-year cycle the stylus arm on the spiral would have been reset by hand and the pointer on the subsidiary dial would automatically reach the next segment to show which third of the Exeligmos cycle the device was displaying.

Translating the glyphs fitted this picture. The Σ stood for ΣΕΛΗΝΗ (Selene), meaning 'Moon', and the H stood for ΗΛΙΟΣ (Helios), meaning 'Sun', and these letters indicated whether a solar or lunar eclipse was due in that month. If both types of eclipse were due in a particular month, the glyph contained both letters. The anchor sign was really a combination of two symbols, Ω and P, meaning 'hour', and the number following it indicated the predicted time of the eclipse after sunrise or sunset. The numbers inscribed in two of the sections of the subsidiary dial – 8 and 16 – indicated that this number of hours had to be added to the predicted eclipse time during that particular Saros cycle.

That was when Freeth knew he had the breakthrough that would make the whole project worthwhile. It was the first evidence that the Greeks had used the Saros cycle in this way. And in Freeth's mind, the discovery completely changed the identity of the device once again. Where Price had presented a calendar computer and Wright had described a planetarium, Freeth saw an eclipse predictor.

Freeth didn't break open the champagne just yet. First, he needed to back up his ideas by deciphering the gears that led to the eclipse dial. The best clue had to be the big turntable with 223 teeth. The machine's maker was unlikely to have bothered with such a large prime number unless he needed it to calculate a particular astronomical ratio. Wright – without the benefit of Fragment F – realised that the number 223 was linked to an eclipse cycle, but had been forced to conclude that the wheel was originally meant for another device. Now Freeth knew that the appearance of the number 223 couldn't be a coincidence. This wheel must have driven the main pointer on the 223-month Saros dial. He worked out that by adding an extra gearwheel next to it (the CT scans showed that there was already a broken shaft in the right place, where such a wheel might have broken off) he could get the turntable to drive the train that led to the lower back dial at just the right speed.

But there were some features that didn't make sense. A 53-tooth gear (another prime number) altered the rotation rate of the 223-tooth gear, only to be exactly cancelled out by another 53-tooth gear on the other side. And there was still the mysterious pin-and-slot mechanism that the 223-tooth gear carried round like a turntable.

It took Freeth six months to come up with an explanation. Like Wright before him, he put together a huge spreadsheet of every possible rotation rate that could be achieved by the gears in this part of the mechanism – allowing for uncertainties in some of the tooth counts and in how some of the wheels were arranged. Then he sifted through the numbers, looking for astronomically significant ratios ranging from various types of month right up to 26,000 years – the period of the Earth's wobble on its axis.

Eventually he realised that the speed at which the turntable was going around – about once every nine years – was the same speed at which the Moon's elliptical orbit shifts around the Earth. Wright had already wondered whether the pin-and-slot mechanism might have had something to do with modelling the variation in the apparent speed of the Moon as it makes its way round its ellipse, but he couldn't fit the idea with the rest of the gearing.

Now Freeth saw how it was done. The wheels in the pin-and-slot mechanism were going round at the speed of the Moon as it circles the Earth, with a wobble in speed corresponding to the near and far points of its elliptical orbit. Meanwhile, this entire cluster was carried round on a slow turntable, once every nine years, modelling the change in orientation of that ellipse around the Earth. But he still couldn't work out how this would have been displayed on the back dials.

Freeth phoned Mike Edmunds to discuss the latest. Edmunds thought for a moment. Couldn't the wobble be sent through to the front of the mechanism, to the lunar pointer, he suggested, so that as it moved around through the zodiac it varied in speed just as the real Moon does? 'No, I don't think so,' said Freeth, since the gear train concerned clearly led to the back of the mechanism. But even as he put down the phone he realised that his friend was right.

The answer lay in the seemingly redundant 53-tooth gears. They enabled the 223-tooth turntable to be used for two different purposes, producing one motion that fed through to the front of the mechanism and one that fed through to the back. The first 53-tooth gear converted the turntable's speed of rotation to match the Moon's shifting ellipse, so that the pin-and-slot mechanism it carried could accurately model the lunar wobble. Once this had been transmitted to the front of the mechanism, however, the second 53-tooth gear reinstated the speed that was needed to drive the Saros eclipse dial on the back.

Michael Wright had been correct that the lunar pointer on the zodiac dial modelled the varying motion of the Moon. But he hadn't needed to imagine an extra epicyclic turntable for this on the front of the device. This part of the mechanism had been there all along.

The revelation was stunning in several ways. Wright had partly predicted it, but Freeth now had direct proof that the device's gears were used to model not just circular motion, but elliptical motion, and a slowly precessing ellipse at that.[4]

4 Today we know that the Moon orbits the Earth in a precessing ellipse, and effectively this is what the Antikythera mechanism is modelling. But of course as discussed in chapter 7, the Greeks themselves wouldn't have seen it this way. They interpreted the varying motions of astronomical bodies they saw as due to different combinations of circles..

The ability it must have taken to think up such a scheme and then execute it was breathtaking – more impressive than a differential gear and beyond any but the most skilled clock-makers today. And the way in which the maker had doubled up the use of the 223-tooth turntable was effortlessly elegant – instead of simply adding gearwheels for extra functions, he had thought through how to strip down the mechanism to the most economic design possible. To mathematicians, the simplest solutions are the most beautiful and this answer – once Freeth grasped it – was everything he had hoped for. Where Price and even Wright had commented on the ancient designer's apparent failings, Freeth now saw that his work made perfect sense.

Beyond the technical ability implied by Freeth's scheme, everything changed too concerning the astronomical knowledge encoded in the mechanism. Whereas the solar and lunar models suggested by Price were pretty basic, and the planetary models suggested by Wright impossible to confirm, the eclipse data and the Moon's wobble would have been state-of-the-art astronomy at the time the device was made. The mechanism was now not just of interest for the history of technology. It became a key piece of evidence in the history of astronomy, encoding the very latest in astronomical knowledge. Looking at the numbers used – the precise eclipse dates recorded, and the exact size of the Moon's wobble – were new clues that could be compared with ancient texts, to tell us more about what Greek astronomers were capable of at the time and to help reveal where the Antikythera mechanism came from.

His quest completed, Freeth hurriedly wrote up the team's results. Wright had published his work in obscure journals read by mechanics and clockmakers – the only audience that mattered to him. But Freeth had much bigger plans. He sent it to *Nature*.[5]

The paper was accepted and scheduled for publication on 29 November 2006. Freeth set about organising a conference in Athens so that on the day that his paper was published he could announce the results to the world. It was to be held in the plush auditorium of the National Bank of Greece, just around the corner from Agamemnon Tselikas's palaeography centre. Freeth invited the whole team, as well as a series of experts in ancient technology and astronomy. He even invited his rival, Michael Wright.

When the day came it was like a carnival, with all of the researchers taken aback by the strength of feeling among the Greek public. Nearly 500 people came to the open session, even though there wasn't room for them in the hall, so the crowd overflowed down the aisles and out of the doors. After Freeth spoke (his words translated into Greek by his colleague Seiradakis), he got a standing ovation that he thought would never end. Every spare moment during the day was taken up by journalists eager for press

5 Mike Edmunds, as holder of a more senior academic position, was officially the paper's 'corresponding author'. However Tony Freeth was the driving force behind the paper, as of the research project.

interviews, and tearful members of the public rushed up to him and the others to shake them by the hand.

After so many lonely years of battling every possible obstacle, Michael Wright found the recognition instantly afforded to Tony Freeth and his team almost too much to bear. He very nearly didn't attend the Athens meeting at all, but in the end he couldn't stay away. Nevertheless his wife Anne stayed at his side to keep him calm. In his talk, later described by one of the other speakers as 'half an hour of continuously controlled rage', he reminded the audience that the latest team was not the only one to work on the mechanism. He started with the story of how, after 20 years studying it, Derek de Solla Price had written to an acquaintance about his final paper on the Antikythera mechanism. 'As far as I am concerned,' Price had written, '[this] wraps the whole thing up.'

'Price seems to have had enough; and I can sympathise,' Wright told the audience, struggling to control his emotions. 'I have studied this artefact for about as long as he did, and whatever pressure he experienced, I can match it. My employer forbade my research, so I have conducted it in my own time and at my own cost, in the face of professional and personal difficulties: intrigue; betrayal; bullying; injury; illness; loss for years of all my data (some still not recovered); the long illness and death of my collaborator; and more.

'Even so . . .' He paused. '*I am still here.*'

He outlined his work, explaining carefully how his radiographs had shown nearly every part of the mechanism's inner workings years before Freeth's team had seen them. He didn't say it in his talk, but he knew that if he had only known about Fragment F he would have got the rest of it as well. He brought his model to show the audience, adjusted to reveal the new arrangement of the lower back dial and lunar mechanism, as Freeth had described it. It took just three hours to make the necessary changes, he said pointedly (see diagram).

On a couple of points he insisted that Freeth was wrong. For example, Freeth played down the possibility that the device showed the planets, despite the inscriptions devoted to this subject. He described the device as an 'eclipse predictor'. But although this emphasised the part of the mechanism that Freeth had discovered, Wright felt it missed the point. It was like finding part of a grandfather clock with a Moon display from which the hour and minute hand had been lost, and announcing that it had been a Moon calculator. Wright remained convinced that the device was primarily a planetarium.

And he stuck to his interpretation that the spiral on the lower back dial had started and finished at the top, not the bottom as Freeth had it. Tony Freeth's moustache bristled when he heard that. This was meant to be a celebration of his own project, yet Wright seemed intent on claiming every discovery for himself. Freeth couldn't resist challenging him when the floor was opened to questions.

'You've got the spiral wrong,' he said. 'We've checked it with the CT, and it starts at the bottom.'

Wright smarted. No matter that the orientation of the

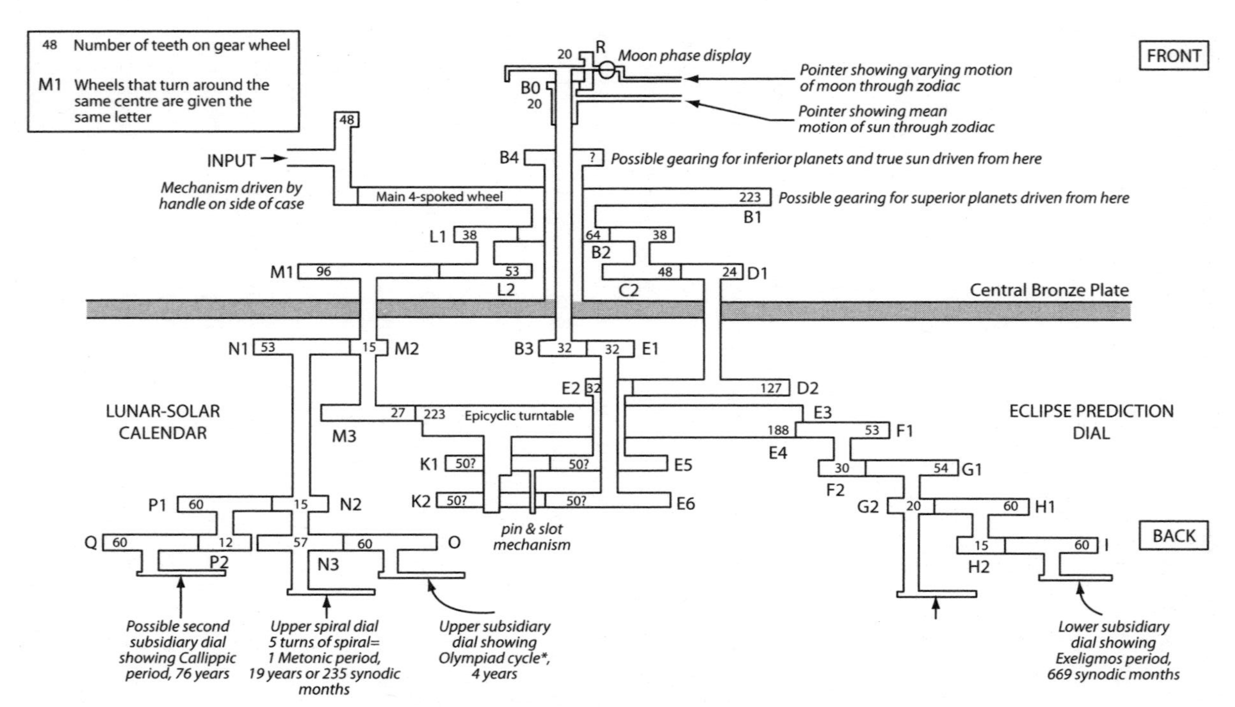

SCHEMATIC DIAGRAM OF THE GEARING INSIDE THE ANTIKYTHERA MECHANISM

* See Chapter 10

spiral was a mere detail. He spoke through clenched teeth, convinced that Freeth was trying to humiliate him.

'I've measured it,' he said. 'And it runs this way round.'

Afterwards, Alexander Jones and John Steele, two experts on ancient astronomy who had attended the conference, came up to Wright.

'Don't take a hacksaw to your model just yet,' they advised. They had noticed something from the slides that Freeth and the others missed: mysterious letters at the bottom of the eclipse glyphs that Freeth hadn't been able to interpret ran alphabetically around the dial, and they showed that Wright was correct about the orientation of the spiral. They were reference letters like those on the parapegma on the front, probably referring to pieces of text around the dial that gave further detailed descriptions of each predicted eclipse.

Then it was time to eat. Wright, Freeth, Roger Hadland, Tom Malzbender, Mike Edmunds, Xenophon Moussas, Yanis Bitsakis and the others all sat down around the same dinner table for one more cry of 'More gears!' Friends or not, they were all part of the same wonderful adventure. As they ate, the news of the incredible Antikythera mechanism sped around the globe. The story zipped down telephone wires, flew through the air, bounced off satellites, inked itself onto newspaper pages and appeared on television and computer screens in every continent of the world. At last, Derek de Solla Price's dream was starting to come true. History was being rewritten, and soon everyone would know that the ancient Greeks had built a clockwork computer.

More than a hundred years after Captain Kontos and his crew raised the Antikythera mechanism from its resting place at the bottom of the sea, the mysterious device had finally been decoded. Whoever turned the handle on the side of its wooden case became master of the cosmos, winding forwards or backwards to see everything about the sky at any chosen moment. Pointers on the front showed the changing positions of the Sun, Moon and planets in the zodiac, the date, as well as the phase of the Moon, while spiral dials on the back showed the month and year according to a combined lunar-solar calendar, and the timing of eclipses. Inscribed text around the front dial revealed which star constellations were rising and setting at each moment, while the writing on the back gave details of the characteristics and location of the predicted eclipses. The mechanism's owner could zoom in on any nearby day – today, tomorrow, last Tuesday – or he could travel far across distant centuries.

For the first time in history it was possible to revisit the past and to predict the future. It was possible to control time itself.

And yet, this can't be the end of the story. Who could have made such a device? And why?

10

Old Man of Syracuse

I know that my day's life is marked for death.
But when I search into the close, revolving spirals of stars,
my feet no longer touch the Earth. Then,
by the side of Zeus himself, I take my share of immortality.

— ATTRIBUTED TO PTOLEMY

THE GREAT OLD circle of Stonehenge on the chalk downs of Wiltshire is one of the world's most awe-inspiring monuments. Its huge stone slabs were hauled into place in various phases during the third millennium BC to serve as a temple dedicated to the worship of the Sun. When the Sun rises on the summer solstice, the longest day in the year, its first rays shine straight down the avenue that leads to the circle and into the open arms of a horseshoe of stones that stands in the centre.

Around 2500 BC the circle's 80 bluestones were lugged to Wiltshire from high in the South Wales mountains. We don't know why the builders undertook such a huge challenge; presumably the stones or their place of origin were of great religious importance. But they weren't alone in their efforts. As the Britons edged their huge load forwards on rafts and rollers, builders working more than 2,000 miles away for the Egyptian pharaoh Khufu were completing their own mammoth project. They cut two million limestone blocks from a quarry in Giza, and piled them high into the most spectacular pyramid ever built.

The north face of the Great Pyramid is aligned almost perfectly to the celestial north pole – the point around which all the stars in the sky appear to spin – with an error of less than a twentieth of a degree. The shafts inside the pyramid may also have been dug to point precisely to the positions of particular constellations in the sky. This is hard to prove, but we do know that many of the Egyptian gods and goddesses were associated with constellations or celestial bodies. The constellation Orion represented Osiris, god of rebirth and the afterlife, while the Milky Way symbolised the sky goddess Nut giving birth to the sun god Re.

In fact, as far back in history as it is possible to look, ancient societies everywhere were obsessed with the heavens. This makes sense from a practical point of view because to a large extent the days and seasons ruled people's lives, but there was a strong spiritual or religious aspect to it too – nearly every culture worshipped gods who lived in the sky. Back then there wasn't much to do at night but look up, and without the electric glare from modern towns and cities the heavenly lights and bursts and trails would have made compelling viewing. Later, observations became more systematic and the ability to track and predict celestial movements would have been a vital part of the transition of any culture from hunter-gatherers into an organised society capable of farming and navigation.

The ancient Greeks were no different and as soon as there are written records we see references to the heavens as a central part of life. In Homer's *Iliad,* written down as early as eighth century BC, the blacksmith god Hephaistos makes a shield for Achilles and decorates it with images of the constellations. And Homer knew that sailors could navigate by the stars – in the *Odyssey* his hero kept the Bear on his left in order to sail to the East. The next surviving work of Greek literature is a poem by Hesiod called *Works and Days*, written around 650 BC. It told farmers what work needed to be done at different times of year, according to which constellations were rising and setting at dawn. These relationships were probably known long before they were written down and would have been remembered as oral poems, perhaps even back to the emergence of agriculture in the region thousands of years before.

Traditionally, the Earth was simply thought of as a flat disc, floating on the ocean like a shield. But by the sixth century BC speculation about the form of the universe was common. Anaximander described a cylindrical Earth suspended in the centre of the cosmos, surrounded by fiery rings. Philolaus, a follower of Pythagoras, suggested that all the celestial bodies including Earth were circling a central fire. The Sun shone with light reflected from this fire, but we didn't see it directly because a counter-Earth (Antichthon) moved around just inside our orbit, always standing in the way. A little later Aristotle adopted the idea of a spherical Earth surrounded by a heavenly realm, with separate circles or spheres carrying round the Sun, Moon, five planets and the stars, which were fixed in place on their sphere like fairy lights poking through cardboard.

In the third century BC an astronomer called Aristarchus worked out that the Sun was many times bigger and heavier than the Earth, and proposed that the Earth must therefore go around the Sun rather than the other way round (he also suggested that day and night were caused by the Earth spinning on its axis). But most of his peers thought it was a terrible idea. For a start, if we were hurtling through space, we'd all fly off. The developing theory of epicycles could account for the movements of the Sun, the Moon and planets around the Earth fairly well, so replacing this with circular motions of the Earth and planets around the Sun didn't make astronomers' models any more accurate (it would take Kepler's elliptical orbits many centuries later to make sense of the heliocentric view).

By the time that the Antikythera mechanism was built – around the beginning of the first century BC – Aristotle's model was more or less accepted, with the wandering motions of the planets explained by epicycles. Astronomers were starting to put numbers to these geometric models to describe the periods of the planets, and the variations in the apparent motions of the Moon and the Sun. The Antikythera mechanism therefore comes from a society for which the nature of the heavens was crucially important, and from a period when astronomers were making their first attempts to describe the universe mathematically. From that point of view it makes sense. But to see these equations converted into a bronze machine is still far beyond anything that we would have expected from the supposedly theoretical Greeks. So what genius came up with the idea, and why?

It is rarely possible to attribute any ancient artefact to a specific individual – unless it happens to be signed. We don't know enough about who was around in the first century BC, and even for those rare characters that we have heard of, we're generally only aware of a tiny fraction of their achievements. It is quite possible – perhaps even likely – that the Antikythera mechanism was dreamed up by someone whose name is forever lost to history.

Bearing this in mind, we can at least speculate based on the clues we have. We know that the wrecked ship on which the Antikythera mechanism was found probably sailed from Pergamon on the Asia Minor coast between 70 and 60 BC, stopping off possibly at Alexandria and almost certainly at Rhodes, on the way to Rome. Pergamon and Alexandria were key centres for trade, and both would likely have boasted the most sophisticated metalworkers around.

But that may not have been enough. The astronomy embodied in the Antikythera mechanism was state of the art, and would presumably have needed the input of a major astronomer. We don't know of any astronomers working in Pergamon at the time. By the first century BC the Romans had taken over the city, so the activity of scientists there may have been waning. Scientific activity at Alexandria was also at a low, after the Roman-friendly king Ptolemy VIII expelled the city's Greek scholars a few decades earlier. Rhodes on the other hand, although its citizens had to be careful not to upset the Romans, was still nominally independent, and one of the few places where Greek academics could work relatively unhindered.

When looking for big names on Rhodes, you don't get much bigger than Hipparchus, one of the most important astronomers of the ancient world. He was born around 190 BC in Nicaea, on the shores of Lake Iznik in what is now Turkey. He moved to Rhodes for the later part of his career, where he made observations in the island's northern hills between 147 and 127 BC.

Hipparchus's work was meticulous and systematic. If he had a purpose in life it seems to have been to persuade other Greek astronomers that their models and theories were useless if they didn't match accurately with observations, and he was not afraid to criticise others if he thought they were wrong. Nearly all of Hipparchus' writings are lost – once the Alexandrian astronomer Ptolemy wrote his epic *Almagest* in the second century AD it superseded all the astronomy that went before (for better or worse), and nothing much else was copied. But we do have one minor work that gives us a glimpse into Hipparchus' character. It is a commentary on one of the most popular poems of antiquity, the *Phaenomena* by Aratus. Like *Works and Days* it listed the rising and setting of constellations through the year, this time along with expected weather patterns. It was much loved for its literary style, but the charm was apparently lost on Hipparchus. He slated it for its lack of accuracy.

Ptolemy based much of his work on that of his Rhodian predecessor, who he described as a 'lover of truth', and we know from the *Almagest* that Hipparchus was one of the first – if not the first – to put numbers to the Greeks' geometrical models of the cosmos. Among other things he compiled the first star catalogue, developed trigonometry, discovered the precession of the constellations through the sky, and may have invented the astrolabe. Of particular interest to us, he was the first to describe mathematically the varying motions of the Moon and Sun, and the pioneering equation he used for the Moon is almost exactly reproduced by the undulating pin-and-slot in the Antikythera mechanism. We don't know of any other astronomer of the time who could have thought of it.

In much of his work Hipparchus was influenced by the precise methods of astronomers from Babylon. They saw any irregularities in the natural world as conveying messages about impending events – usually bad ones. This included the births of deformed animals or people, animals behaving oddly or strangely formed plants. But the most important sources of such messages were the heavenly bodies, and Babylon's astronomer-priests kept them under close surveillance. It was like watching a soap opera set in the sky – the Milky Way was known as the 'river of heaven' and the erratic planets were seen as gods riding about the celestial countryside as living men travel on Earth.

Similar beliefs were held across ancient Mesopotamia – a broad area between the Tigris and Euphrates rivers, roughly corresponding to modern-day Iraq, and parts of Turkey, Syria and Iran. The Babylonian empire was in the south of this region, while Assyria was in the north. Much of what we know about these omens comes from a series of Babylonian clay tablets called *Enuma Anu Enlil*, which contained lists of astronomical events and the omens derived from them. The tablets were discovered in the library of the Assyrian King Ashurbanipal in Nineveh on the river Tigris (near the modern city of Mosul in Iraq). Ashurbanipal reigned in the seventh century BC, but he collected old cuneiform texts from all over Mesopotamia and especially Babylon, and the omens themselves are thought to date back to the second millennium BC. The timings of the Moon's phases, risings and settings of the planets and especially eclipses all bore messages about the well-being or otherwise of the king and his country. Some predicted floods or war or the quality of the coming harvest; others were more specific about the personal fate of the king. A typical example reads: 'When in the month Ajaru, during the evening watch, the moon eclipses, the king will die. The sons of the king will vie for the throne of their father, but will not sit on it.'

The court astronomers concerned with monitoring these omens made daily observations of the state of the sky, so that they could keep their king informed of his impending fortune. This was crucial, because once a sign was spotted it was possible to avert the worst by carrying out the appropriate rituals. Often this involved performing lamentations or making offerings to the gods, but sometimes more drastic action was called for. One ritual involved taking a beggar from the street and sitting him on the throne for the duration of a lunar eclipse, so that the divined ill-fortune would befall him and not the temporarily abdicated king.

When the omens were first compiled, the astronomers probably observed each celestial sign directly, then enacted the appropriate ritual. But after a few centuries of keeping meticulous records they started to establish patterns in the recurrence of different heavenly events, until eventually they didn't need to see them at all and could predict them in advance, even the seemingly aimless motions of the planets.

Babylonia and Assyria became part of the Greek world in the fourth century BC when they were conquered by Alexander the Great. After that there were no indigenous kings for astronomers to report to, but their cults remained an important part of life. Babylon gradually emptied as the Greek rulers moved the locals to their new capital Selencia on the Tigris, but a small group of priests held out at the city's temple complex. For another couple of hundred years these lonely astronomers carried on observing the skies and doggedly tended statues of their gods as if they were living beings – feeding them, clothing them and parading them around the empty temple.

Isolated snippets of information from the Babylonian

astronomers had already started to filter through to the Greeks. But during the second century BC Hipparchus took this much further, basing large sections of his work on their data. We know from Ptolemy that Hipparchus relied on a list of eclipse observations reaching back centuries and that he converted long stretches of planetary observations into the Egyptian calendar (the same one represented on the front of the Antikythera mechanism), so that they were more convenient for Greek astronomers to use. The Babylonians were the only ones we know of who had such data. The historian of astronomy Gerald Toomer, who studied Ptolemy's writings extensively, concluded that Hipparchus must have travelled to the temple in Babylon himself and worked with the last astronomers there to extract the information that he needed from the old tablets.

Seeing the accuracy that these priests achieved is probably what inspired Hipparchus to strive for similar precision in the geometrical models that his peers were throwing about at the time. He used the Babylonian data to derive numbers for the models and in doing so transformed Greek astronomy from a largely theoretical science into a practical, predictive one. The reason that Hipparchus is such an important figure is because in him two long traditions of astronomy became one. When Ptolemy later built on Hipparchus' work it set the foundation for all of western astronomy until Copernicus and Kepler kicked out the geocentric view and reinstated the Sun at the centre of things in the sixteenth and seventeenth centuries.

It is tempting to think that the Antikythera mechanism – a machine that also embodies both the geometrical circles of the Greeks and the precise arithmetic of the Babylonians – might have had something to do with Hipparchus. If you take the very earliest end of the date range suggested by the inscriptions, then it's just possible that he was still alive on Rhodes when the mechanism was built. So, if the device was made by Hipparchus or his followers, what would they have used it for?

The mechanism displayed the state of the skies at any chosen moment in time. It incorporated sophisticated astronomy theory and was clearly made by someone who cared a lot about making it as accurate as possible. But it wasn't an astronomer's workshop tool. There's no obvious way in which the device could have been used to make observations. And it was covered in idiot-proof inscriptions, so it probably wasn't intended for use by someone with specialist knowledge.

One possible purpose is to cast horoscopes. The art of astrology – the idea that a person's fate depends on the configuration of stars and planets at the time of their birth – was just becoming popular in the Hellenistic world at the time that the Antikythera mechanism was made. A key requirement for practising astrology is being able to work out what the stars and planets were doing when a person was born (otherwise you'd always need an astronomer on hand at the birth to note it all down just as the baby pops out). The Greeks couldn't do this for the planets before the second century BC. But when Hipparchus brought back the Babylonian's arithmetical models of the planets' motion – making it possible to use written tables to extrapolate backwards or forwards in time – astrology took off. If the Antikythera mechanism showed the planets, it would have been the ultimate luxury gadget for checking the state of the skies at any desired moment, something wealthy clients (including the Romans, who loved astrology) would have paid a lot of money for.

We don't have any writings from Hipparchus on astrology, but he seems to have advocated the idea – the heavens governed the seasons and the tides, after all, so why shouldn't they influence other things on Earth too? For example, the Roman historian Pliny the Elder wrote in the first century AD about astrology's great debt to Hipparchus, saying that he 'can never be sufficiently praised, no one having done more to prove that man is related to the stars and that our souls are part of heaven'.

There is another big name from Rhodes, however, who suggests a rather different interpretation of the Antikythera mechanism. Suspect number two is a philosopher called Posidonius, and he was working on the island at exactly the time our clockwork box set off on its last journey. He was born in 135 BC in Apamea, a Greek city on the river Orontes in Syria. He settled in Rhodes around 95 BC, where he founded a school. He was nicknamed 'the champion' and seems to have been universally recognised as one of the most wise and learned men in the Hellenistic world.

Posidonius was also a senior politician – he was president of Rhodes for one six-month term, and he travelled to Rome as Rhodes' ambassador in 87–86 BC, around the time that Sulla was smashing his way through Athens. He made friends with some key Roman figures there, including Pompey, the up-and-coming general. Pompey visited Posidonius on Rhodes in 66 BC, just before he set off to take on the stubborn king Mithridates (he asked the teacher for advice, who replied diplomatically: 'Be ever the best') and again on his triumphant return. According to Pliny, Pompey stopped his assistant from striking on Posidonius's door, and instead 'the man to whom the East had bowed in submission bowed his standard before the door of learning'.

As if that wasn't enough, Posidonius excelled in geography and science too. He travelled widely before settling in Rhodes – to Greece, Spain, Africa, Italy and most famously to the Celtic lands of Gaul, in the decades before the Romans took over. He wrote vivid descriptions of the wild and violent goings on he found there, including how the warriors would hang the severed heads of their enemies in doorways when they got home (something the scholarly Posidonius found nauseating at first, though he got used to it after a while), and how at feasts, which often ended up in rowdy fights to the death, a man would sometimes take pledges of gold or wine, distribute it to his friends, then lie face up on his shield so that someone could slit his throat with a sword for the general amusement of the party.

On his travels he observed the regular patterns of the tides, and theorised that they were caused by the motions of the Moon. And back in Rhodes he used astronomical observations to estimate the distance and size of the Sun. He was a key member of the Stoic school of philosophy and his interest in the heavens came very much from his Stoic world view, which interpreted the cosmos as a single organism, infused with a divine, intelligent life force that gave it form and direction. Could Posidonius have been behind the Antikythera mechanism?

If so, it was probably meant not as an astrological tool, but as a philosophical or religious demonstration of the workings of the heavens. We have an important piece of evidence to support this idea, from the Roman lawyer and politician Cicero. He spent time on Rhodes when he was a young man in the early 70s BC, just a few years before the Antikythera ship sailed. (His trip was possibly motivated by a desire to keep out of General Sulla's way, because Cicero had just implicated an associate of his in a murder case.) While in Rhodes, Cicero studied with Posidonius and later wrote about an instrument 'recently constructed by our friend Posidonius, which at each revolution reproduces the same motions of the Sun, the Moon and the five planets that take place in the heavens each day and night'.

Cicero went on to ask: 'Suppose someone carried this to Scythia or to Britain. Surely no one in those barbaric regions would doubt that the orrery had been constructed by a rational process?' His point was that just as the sphere had an intelligent creator, so did the universe. Posidonius believed more in a guiding life force than a distinct creator, but to philosophers of all persuasions a machine that modelled the workings of the solar system would have been a powerful reassurance of the order or purpose that lay behind the nature of things.

Despite having a name that meant 'chickpea' (thanks to an ancestor whose nose apparently resembled one), Cicero was pretty much the leading intelligence in Rome for his time. He was one of the first people to write about philosophy in Latin – his mission was to make Greek teachings available to the Roman reading public. But his political judgement was often flawed. While trying to play in Rome's big league, for instance, he got caught between the generals vying for leadership (including Julius Caesar and Pompey) and in the end was hunted down and killed by Mark Antony's men in 43 BC.

Although Cicero's writings are generally trusted, scholars throughout history took his account of Posidonius' device with a large pinch of salt. Cicero had no scientific training and he provided no technical description of how the instrument worked. To simulate the movements of the planets would have taken a sophisticated system of gears, which was thought way beyond the craftsmen of the time, so the whole things sounded pretty unlikely. After all, he could have made the story up just to illustrate a point. However, the Antikythera fragments prove that the technology did exist and suggest that Cicero's description can be taken seriously after all. Perhaps our mechanism was commissioned by Posidonius as a demonstration piece for all who came to study there, or as a prized gift for an important visitor, such as Pompey.

Modelling the heavens with geared devices ran alongside a parallel philosophical tradition of modelling living creatures: people, animals and birds. These didn't use gearwheels, instead they were powered by steam, hot air and water. The tradition seems to have started with the engineer Ctesibius, working in Alexandria in the third century BC. He was a big specialist in water clocks, many of which included automated figures. The Tower of the Winds, reconstructed by Price according to one of Ctesibius's designs, may also have incorporated little puppets that moved alongside the turning disc of the heavens.

The engineer Hero, who worked in Alexandria in the first century AD, built on Ctesibius's work. He invented the first steam engine, as well as the first vending machine, which dispensed holy water, and also the first wind-operated organ. He created all sorts of wondrous mechanical shows, both for theatres and for temples. In one example, a platform carrying a figure of Bacchus (the god of wine and intoxication) moved forwards to an altar causing a flame to burst forth, while wine flowed from his cup and automated figures danced around the temple to the sound of drums and cymbals.

Historians have often scoffed at the Greeks for wasting their technology on toys rather than doing anything useful with it. If they had the steam engine why not use it to do work? Looking at the Antikythera mechanism we might also ask: if they had clockwork, why not build clocks? Many centuries later, in Europe, such technology prompted the Industrial Revolution, ushering in our automated modern world. Why did it not do the same for the Greeks?

In the 1960s and 70s the popular view was that because the Greeks had slaves to do manual labour there was no incentive to develop technology to replace them. That argument has now fallen out of fashion and most experts will tell you the answer is more complicated than that. Even in slave-owning societies machines can benefit owners by allowing slaves to do more work than they did before – such as when the cotton gin caught on in the plantations of the American Deep South at the end of the eighteenth century.

It probably has more to do with what the Greeks would have regarded as useful. Models of people and animals, like those of the cosmos, affirmed the idea of a divine order. The Greeks didn't think that real creatures ran on steam, just as they didn't think the planets were powered by cogs, but the fact that they clearly had a designer argued implicitly that the same was true for the universe. Gadgets like Hero's also went beyond that general principle to demonstrate basic physical laws in pneumatics and hydraulics.

Take Hero's famous steam engine, for example. This consisted of an airtight ball-shaped chamber mounted on to a horizontal metal axis. The axis was also a pipe that fed steam into the sphere as a water chamber beneath was heated, and the only way that the steam could escape was through two bent nozzles, one on either side of the chamber. This set-up forced the steam out of the ball perpendicular to its axis, causing it to spin around at high speed.

There is no obvious way that this contraption could be made to do useful work, and it has often been derided as a mere toy. But a more recent interpretation is that Hero intended his engine to demonstrate a particular physical principle. Several centuries earlier the philosopher Aristotle had argued that for any animal to move it must be supported on something 'unmoved and resisting'. A walking man pushes against the earth, for example, and a fish pushes against the sea as it swims. But a man in a boat won't go anywhere however hard he pushes against the inside of his vessel. Aristotle also applied this principle to the spheres that he believed carried the celestial bodies. He argued that their circular motions were passed on to them by the outermost sphere, which in turn derived its motion from the 'unmoved mover', in other words, God.

It has been suggested that Hero built his steam engine to disprove Aristotle's theory, as the force of its motion derived from within the sphere, with no need for an external source. As with the Antikythera mechanism, this 'toy' was far from trivial. The aim was to advance a philosophical principle, to aid understanding of the universe and improve oneself in the process. To what better use could technology be put?

There was a practical side to this as well. As a member of the relatively small ruling elite in the Hellenistic world, the most useful thing you could do in terms of gaining power and prestige was to impress your peers. You did that not by timing an hour more accurately or ploughing a field faster, but by demonstrating what you knew.

This wasn't about education so much as inspiring awe and wonder. In many of Hero's devices, as with the Tower of the Winds, the working mechanisms were hidden – it was all about putting on a show.

For example, Hero described a mechanism that would secretly channel hot air from a fire on an altar and use it to open the temple doors. There was also an automatic theatre featuring thunder and fire, and a mirror that showed a goddess in place of the viewer's reflection. Some scholars argue that Hero's *baroulkos* – the geared machine for lifting heavy weights – was inspired as much by the trick of allowing a man to lift hundreds of times his weight as by the prospect of using it to do work.

Inspiring wonder in the masses by having metal animals come to life or temple doors mysteriously open played a role in keeping the lower classes in their place and stabilising the social order. Having knowledge and understanding of how the world worked was part of being in the Greek ruling elite. It also helped to impress the visiting Romans, who had an appetite for the Greeks' scientific instruments as well as their art.

We know much less about the range of planetary mechanisms built than we do about Hero's models, and if it wasn't for the Antikythera fragments we wouldn't be sure that they existed at all. But as well as fitting within a similar philosophical tradition, such devices may have been equally widespread.

The first argument for this relies on simple statistics. We have a very distorted view of ancient documents, because they only tended to survive if scribes throughout history thought they were worth copying. Unfortunately, this doesn't mean that the best stuff survived – often quite the opposite. If intellectual standards dropped over time, for example, scribes didn't copy the texts that represented the height of earlier scholarship because they couldn't understand them; instead they chose simpler works with popular appeal. Imagine some future civilisation judging our scientific knowledge based purely on hints from *Friends* and *Big Brother*.

With objects the situation is even worse, especially for anything made from valuable material like bronze, which, unless out of reach at the bottom of the sea, was inevitably melted down and recycled. Only a tiny fraction of what was made survives. For example, across the Greek world we know that there were hundreds of thousands – if not millions – of large bronze statues. Pliny wrote that there were 3,000 in the streets of Rhodes city alone, and this was in the first century AD, when the Roman-occupied island was a poor shadow of its former gleaming self. In the National Archaeological Museum in Athens, which has one of the best collections of Greek bronze statues in the world, there are now just ten. All but one are from shipwrecks.

So in terms of the Antikythera mechanism, the fact that we have found even one suggests that it can't have been unique. We can't say that such devices were ever mass produced, but there could have been dozens if not hundreds of them.

The sophistication of the mechanism supports this idea. This was not the work of a novice craftsman trying out his skills with clockwork for the first time. It would have taken practice, and once someone had the idea of using gearwheels to simulate the heavens, the design would probably have taken generations to perfect. The components of the mechanism are also very small – about as small as you could make them without needing eye glasses (which, as far as we know, the ancient Greeks didn't have). The design and the maker's skills must have been honed on a number of simpler, bigger mechanisms before being scaled down.

Finally, we have more clues from texts, which tell us that Posidonius wasn't the only thinker linked to such a device. Cicero and several other Roman writers told of a similar instrument built by the great Archimedes, who worked in Syracuse, Sicily, in the third century BC. Cicero described it as a sphere that showed the motions of the Sun, Moon and planets around the Earth. 'The invention of Archimedes deserved special admiration because he had thought out a way to represent accurately by a single device for turning the globe those various and divergent movements with their different rates of speed,' he wrote. 'The Moon was always as many revolutions behind the Sun on the bronze contrivance as would agree with the number of days it was behind it in the sky.'

Claudian, writing in Rome in 400 AD, was more poetic: 'An old man of Syracuse has imitated on Earth the laws of the heavens, the order of nature, and the ordinances of the gods. Some hidden influence within the sphere directs the various courses of the stars and actuates the lifelike mass with definite motions. A false zodiac runs through a year of its own, and a toy moon waxes and wanes month by month. Now bold invention rejoices to make its own heaven revolve and sets the stars in motion by human wit.'

The device must have been highly prized, because the Roman general Marcellus took it home with him in 212 BC when his army sacked Syracuse, killing Archimedes in the process (supposedly as he was in the middle of a mathematical proof). The globe was the only thing Marcellus claimed for himself from the huge booty that was captured, and it was kept in his family in Rome for generations, until Cicero saw it years later. He wrote that Archimedes must have been 'endowed with greater genius than one would imagine it possible for a human being to possess' to have built such an unprecedented machine. Cicero was a big fan of Archimedes; while serving as Rome's representive in western Sicily in 75 BC he had sought out the mathematician's grave, which he found covered in brambles and thorns, and cleaned it up as a mark of respect.

Again, historians have never known quite what to make

of such descriptions, none of which includes any technical details about how the device was made. But the Antikythera mechanism helps us to take the story seriously. The way that Cicero talks about Archimedes ('endowed with greater genius than one would imagine it possible') suggests that he was the first to build such a contraption. Whoever was behind the Antikythera mechanism itself, the tradition may well have started generations earlier, with Archimedes.

The theory of epicycles was very new when Archimedes lived, if it existed at all, and there was certainly no way to model the elliptical orbits of the Moon and Sun. So his device might have been relatively simple, perhaps a schematic model showing the Sun, Moon and planets rotating around the Earth at various but constant speeds. Cicero simply said that Archimedes 'made one revolution of the sphere control several movements utterly unlike in slowness and speed', so it's possible that this was the case. Later, other engineers could have built on the tradition, coming up with more sophisticated gearwork to incorporate the latest astronomical knowledge – including that of Hipparchus – as it became available. Perhaps Hipparchus or his work influenced a switch on Rhodes from a schematic model to a mathematical calculator that displayed the precise timing of celestial positions and events on its dials.

It's impossible to know for sure, of course. But we know that Archimedes pioneered the use of gears in simple weight-lifting devices, using single pairs of differently sized wheels to change the force applied to an object. Maybe he also had the idea of using more precise clockwork to control the speed of spinning model planets. One of the few biographical details that slipped into his treatises was that his father Phidias was an astronomer, so it makes sense that Archimedes would have been interested in the heavens. We also know that he spent time working with Ctesibius in Alexandria before he moved to Syracuse, so perhaps the seeds of both modelling traditions – of planets and of living creatures – were sown with the pair of them there. Intriguingly, the mathematician Pappus, working in Alexandria in the fourth century AD, said that Archimedes wrote a treatise called 'on sphere-making', apparently his only work on 'practical matters'. No copy of this document survives, but it's not a huge stretch to imagine that it explained how to build devices that model the movements of celestial bodies around the Earth.

More recent studies of the Antikythera mechanism suggest an even stronger link with Archimedes. After Tony Freeth's *Nature* paper was published in 2006 he called in Alexander Jones, a historian of astronomy now working at the Institute for the Study of the Ancient World in New York. Jones worked with Yanis Bitsakis and Tony Freeth to make a closer study of the Greek inscriptions that had been revealed in X-Tek's CT images and in particular the letters covering the five-ring spiral dial on the upper part of the instrument's back face. Michael Wright had originally shown that this spiral was divided into 235 sections, depicting the 235 synodic months of the 19-year Metonic cycle, which tracks the motions of the Sun and the Moon.

The results, published in *Nature* in July 2008, were completely unexpected. One surprise came from the subsidiary dial located inside the main spiral. The gear train leading to it is lost, but because it was divided into four both Michael Wright and Tony Freeth had assumed that it represented the Callippic 76-year cycle – four times the 19-year cycle displayed on the main spiral. But when Jones read the names inscribed on this dial he realised that it was doing something quite different. The inscriptions – Isthmia, Olympia, Nemea and Pythia – referred to the Panhellenic games, at which athletes from across the Greek world gathered to compete in events such as running, long jump, discus and wrestling.

Knowing which games were held when was of no astronomical use, but it had huge cultural significance. The Greeks often kept track of time by using the Olympiad 4-year cycle, so this dial would have enabled the user of the Antikythera mechanism to convert the date shown on the front dial into the Olympiad calendar. The presence of this dial supports the idea that the mechanism was not an astronomer's tool but was meant for popular demonstrations, albeit to relatively small groups of educated intellectuals (the mechanism would have been too small to display to a large crowd).

Alexander Jones was also able to read the month names inscribed on the surviving sections of the main spiral, and he found that they too came from a local civic calendar. The inscriptions showed which months should have 29 days instead of 30, as well as which years should have 13 months instead of 12, so that the calendar fitted neatly into the 19-year astronomical cycle also shown on the dial. The calendar follows rules similar to those described by the astronomer Geminus, who worked on Rhodes in the first century BC, and the spiral was arranged so that the 29-day months all lined up along the same spokes.

The discovery of this civic calendar raised an exciting possibility. In ancient Greece, different cities used different sequences of months in their calendars, so it was now possible to investigate where the month names on the Antikythera mechanism came from. Jones did indeed find a match – and it turned upside down everyone's ideas about where the mechanism was made. The month names inscribed on the device do not come from Rhodes. Instead, they were used in colonies founded by the city state of Corinth in central Greece. Not much is known about the calendar used in Corinth itself, but the month names used on the Antikythera mechanism are similar to those used in Illyria and Epirus in northwestern Greece, and in Corfu – all of them Corinthian colonies. Corinth had another important colony, however: Syracuse, where Archimedes lived. We don't have any direct evidence of the calendar used in Syracuse, but the closest match of all to the month names on the Antikythera mechanism is with the calendar of Tauromenion in Sicily, which is thought to have been founded by settlers from Syracuse. Seven of the months on the mechanism are identical in name and sequence to those used in Tauromenion, and it seems likely that the original settlers took these directly from the calendar of their home state Syracuse.

Corinth and Epirus were destroyed by the Romans in the second century BC, so it's unlikely that the Antikythera mechanism – made several decades later – was created for use there. But Syracuse, despite having been sacked by Marcellus in 212 BC, was still Greek-speaking in the first century BC and relatively prosperous. The Romans exacted heavy taxes, but as in Rhodes, the city's citizens were reasonably free to get on with their lives. So, this new evidence suggests that although the Antikythera mechanism almost certainly began its final voyage in the eastern Mediterranean, it was originally made by (or for) someone in Syracuse in the west.

The Antikythera ship did not stop at Sicily – it sank when it was still far to the east of that island. But it was headed in that direction; its likely route to Rome would have taken it straight past Syracuse. Perhaps the wealthy owner of the Antikythera mechanism had visited Posidonius' school in Rhodes to show off his latest toy to the philosophers there, then boarded the ill-fated vessel on his way home. Or perhaps the mechanism was made to order by one of Rhodes' finest craftsmen and was being delivered to a buyer in Syracuse. However, the dating of around 100 BC suggests that the instrument was probably several decades old when the Antikythera ship sailed in 70–60 BC. So its owner could have moved from Syracuse to Rhodes or elsewhere in the eastern Mediterranean and taken the mechanism with him. Or perhaps it was taken east as a prized gift or religious offering. Later the device ended up being carried west again, as booty for Rome.

Whatever the Antikythera mechanism's story, it seems likely from all the evidence, including Cicero's writings, that similar geared models were being made at this time in both Syracuse and Rhodes. The mechanical tradition begun by Archimedes in Syracuse a century earlier was still going strong, with his original design being updated by the latest astronomical knowledge from Rhodes and elsewhere as it became available. The latest models were then shipped across the Greek-speaking world.

In fact, a wide tradition of such devices seems to have continued until at least the fourth century AD. The mathematician Pappus, who lived in Alexandria then, wrote that in his time there was a whole class of mechanics called sphere-makers who 'construct models of the heavens'. These models may never have got any more sophisticated than the Antikythera mechanism, however. Developing complex technology takes a thriving urban environment with stability, money, skilled craftsmen and rich clients. All those could be found in the Hellenistic world, but it wasn't to last for long. By the beginning of the first century BC, Syracuse and Rhodes were two of the last places that Greek scholars could work uninterrupted. But Syracuse gradually declined under the influence of the Romans and Rhodes too was sacked by the Roman general Cassius in 43 BC, and never regained its former greatness.

Although they appreciated Greek science and philosophy

as much as Greek art, the Romans never practised much science themselves, and there was a gradual decline in scientific learning throughout the period of the Roman empire. From the third century AD onwards the few scientists that were around did little original work; instead, they just tended to write commentaries on the works of their Hellenistic predecessors (Pappus was one of the last great Greek mathematicians). And when the Roman empire fell, the light of scholarship in Europe went out almost completely. It took nearly a thousand years for western society to recover.

We now know, of course, that Price was right when he argued that the technology incorporated in the Antikythera mechanism wasn't totally lost. The sixth-century geared sundial that was brought in pieces to Judith Field and Michael Wright at the Science Museum proved to be a vital piece of evidence, showing that at least a simplified form of the technology survived into the Byzantine Empire. (So did the tradition of automated figures. In the tenth century, Emperor Constantine VII was still religiously following the principle of inspiring wonder in the masses – Bishop Liutprand of Cremona wrote after visiting him that his throne, which could be raised and lowered at will, was surrounded by mechanical beasts, including roaring lions and a tree with singing birds.)

During the seventh and eighth centuries the Arabs conquered large regions, including Syria, Iraq, Egypt, Mesopotamia, Iran and Spain. The rulers converted their new subjects to Islam and saw it as their duty to make the old Greek knowledge available in Arabic. During the ninth century they funded efforts to translate all the Greek scientific texts that could be found, even going on missions into Byzantine territory to rescue them. Price knew of two examples of Islamic geared calendars – the 'Box for the Moon' described by al-Biruni in Ghazna (now in Afghanistan) in the eleventh century, and one attached to a surviving astrolabe made in Isfahan (now in Iran) in the thirteenth century.

Another example has turned up recently – the Box for the Moon is also described in a tenth-century treatise that appeared for sale in 2005, attached to a sundial and with a layout exactly like the one that Michael Wright reconstructed for the Byzantine instrument. The treatise isn't signed, but is thought to have been written by an astronomer called Nastulus, who worked in Baghdad around 900. The fact that this version of the geared calendar is attached to a sundial, as in Wright's instrument, makes the link between the Byzantine and Islamic instruments even more certain, and together all of the examples suggest that the idea of using gears to model the motions of the Sun and Moon was common in the Islamic world and inherited directly from the Greeks.

Islamic engineers also built on the Greek tradition of water clocks and they described impressive timepieces driven by both water and mercury, including one called the 'clock of Archimedes'. Some of them had rotating dials to represent the sky, just as Derek de Solla Price described for the Tower of the Winds, as well as moving figures and audible chimes, such as balls dropping on to cymbals. Most had very simple gearing, however. They were limited because in general running water isn't powerful enough to drive large numbers of wheels (for this reason the Antikythera mechanism was almost certainly turned by hand).

But there's one exception that supports Price's idea that Greek gearing techniques used in the Antikythera mechanism directly influenced the development of clocks. It's an Arabic manuscript, only discovered in the 1970s, but written in the tenth or eleventh century in Andalusia by an engineer called Al-Muradi. In it he describes a number of water clocks. The manuscript is badly defaced, but it's just possible to get an idea of how they work. Most water clocks we know about from the Islamic world are quite delicate devices, but Al-Muradi's are large and rugged – driven by fast-moving streams and involving huge wheels, ropes and hefty weights. The gearing gets quite complex, including what look like epicyclic gears. Al-Muradi said he was writing to revive a subject in danger of being forgotten, suggesting that rather than being a new invention the technology had been around for some time. So it's possible that epicyclic gearing, too, made it through from the Greeks.

Meanwhile, the wheel of history kept turning. Europe had been gathering its strength and in a series of Crusades spanning the twelfth and thirteenth centuries much territory was won back from the Muslims, including Spain. Again, the new rulers wanted to make the old knowledge available to the Christian world, and they funded a systematic effort to translate old documents into Latin – both Arabic versions and the Greek originals.

By this time the Catholic church was keenly interested in finding a means for accurate timekeeping to control the work and prayers of its monks. It had been using marked candles for centuries, but as soon as Islamic knowledge started to make it into Christian Europe water clocks appeared in monasteries there. In 1198, during a fire at the abbey of Bury St Edmunds, the monks 'ran to the clock' to fetch water. And an illustration in a manuscript dated to around 1285 shows a water clock in a monastery in northern France, with a wheel that rang bells as it turned.

Around this time, some unknown genius finally invented the one component that was still needed to make the switch to fully mechanical clocks: the escapement. It was particularly important for the monasteries to have clocks that rang bells (to make sure that the monks woke up for prayers during the night). Perhaps someone was experimenting with getting oscillating weights or hammers to strike bells as an alarm mechanism. Instead of being driven by the clock, they realised that the oscillating motion allowed the weights to regulate the power needed to drive it.

Once clocks were mechanically driven they were powerful enough to incorporate many more gearwheels and they exploded in complexity. Within a few decades clocks appeared all over Europe and almost immediately they incorporated elaborate astronomical displays with dials and pointers that were spookily similar to their ancient predecessors, such as the Antikythera mechanism and the Tower of the Winds. The clock of Richard of Wallingford, completed in St Albans in 1336, is one of the first that we know of. It had a large astrolabe-style dial that showed the position of the Sun in the zodiac, the Moon's age, phase and node, a star map and possibly the planets. (Modern additions included a Wheel of Fortune and an indicator of the state of the tide at London Bridge.) Giovanni de' Dondi completed his clock in Padua, Italy, in 1364. As the reconstruction in the Science Museum still shows, this seven-sided construction had dials showing the time of day, the motions of all the known planets, a calendar of fixed and moveable feasts, and an eclipse prediction hand that rotated once every 18 years according to the Saros cycle.

The speed with which these astronomical displays became so elaborate and so widespread – and their similarity to those developed by the Greeks – suggests that all this did not emerge from scratch. Various pieces of the necessary technology, and the idea of using gearwheels to simulate the heavens, must have been lying dormant in a range of devices – including water clocks and hand-driven calendars – so that when the invention of the escapement allowed the construction of a mechanical clock, all of these old tricks came rushing out of the wings into the new tradition.

Once again, geared astronomical displays were used to demonstrate the wonder of the heavens – and to solidify the position of the church – and they were placed prominently in big clock towers and public squares. They often included mechanical figures, too, clearly influenced by Greek automata. The St Mark's clock in Venice, completed in around

1500, which used seven concentric dials to display the time and the movements of the Sun and Moon in the zodiac, was topped by two bronze giants that struck an enormous bell, as well as moving statues of Mary, Jesus and the Three Kings. And the Strasbourg clock, completed around 1350, included a bronze cock that flapped its wings at midday and crowed three times through a small organ in its throat.

Leonardo da Vinci, in particular, was obsessed by Hero's work and studied everything he could find on it through Arabic translations. He developed clocks equipped with figures for striking the hours, as well as flying birds, and once, for the French King Francis I, he built a lion powered 'by force of wheels' that walked along before opening its chest to display a bouquet of flowers.

As well as providing the technology and skills that ultimately helped to trigger the Industrial Revolution, these mechanical devices were important in changing how people thought about the universe. Instead of representing an animate cosmos ruled by a guiding life force, scientists started to think of an inert, mechanistic universe that followed natural physical laws. The Antikythera mechanism was originally meant as a celebration of the heavens. But as clocks developed, the ability to measure minutes, seconds and even shorter time periods finally broke our ties with the sky. We're free, like no other civilisation has been, from the cycles of the heavens – the first people to be ruled by our watches and not by the Sun.

★

One mild November evening in 2006 – just before the excitement of the conference at which they were to present their solution of the Antikythera mechanism – Yanis Bitsakis and Xenophon Moussas treated me to dinner at a little restaurant a couple of blocks from the Athens National Archaeological Museum. Over aubergine and octopus they told me about the strange hold that the Antikythera mechanism exerts over whoever gets close to it and how one day they would like to devote an entire museum to the story of the fragments and those who have studied them. 'It represents the same way we would do things today; it's like modern technology,' said Bitsakis. 'That's why it fascinates people.'

He's right of course. Setting eyes on the fragments is an intoxicating experience, precisely because the secrets they hold are so familiar. They give us an unprecedented glimpse into another time, into people who thought like us, solved problems like us, built machines like us. You can see instantly that there, in those flaky green fragments, are the seeds of our entire modern world.

However, my lasting impression is not of the similarities between our world and theirs, but the differences. We now understand more about the universe than any previous civilisation could have dreamed. We observe and measure the objects in our solar system in exquisite detail, calculate and predict their movements by the nanosecond and send spacecraft to visit them. We have photographed the Earth from space, sent men to the Moon and beamed pictures back from Mars. We have caught stardust from the tail of a comet and probed the atmospheres of planets circling distant suns. We understand better than ever the true extent of our universe, how it began and how it will end, and the nature of our place within it.

Have we also lost something? At the very least, we're missing out on the best light show on the planet. Living in today's permanently illuminated towns and cities, most of us have little sense of the rhythms of the sky; the intricate dance of the Earth, Moon and Sun, the wanderings of the planets or the circles of the stars. Finding out who made the Antikythera mechanism and why also turns upside down any notion we might have had about ancient technology being 'primitive' and our own being so 'advanced'. After all, where we see practical machinery that can measure time accurately and do work, the Greeks saw a way to gain knowledge, demonstrate the beauty of the heavens and get closer to the gods.

Epilogue

BACK IN ATHENS, the Antikythera fragments are still hanging in their glass case. They look as corroded and battered as ever, but their story is out at last.

The work to interpret them goes on, however. Alexander Jones is still working on deciphering the inscriptions on the mechanism and he believes there are many more insights to come. John Steele of the University of Durham, an expert in ancient eclipse prediction, has also been working with Tony Freeth to read the inscriptions on the lower back dial. They have now studied 18 eclipse glyphs in total and confirmed that they were labelled with alphabetical letters that would have referred to text elsewhere on the mechanism, giving further details about each predicted eclipse. The details were published alongside Jones' findings in *Nature* in July 2008.

As well as the letters indicating whether each event was a lunar eclipse or a solar eclipse, the glyphs were labelled with either 'H' or 'N', to indicate whether the eclipse occurred during the night or during the day. Finally, the researchers worked out why the spiral of the eclipse dial had four rings. The speed of the Moon as seen from Earth fluctuates in a cycle that lasts 14 months, and the dial was arranged so that each quarter was divided into exactly 14 sections. This ensured that whichever direction the dial's arm was pointing in, the speed of the Moon (one of the factors determining the duration of an eclipse) was the same across each of the four rings of the spiral.

Roger Hadland's gamble that the Antikythera project would turn his company around paid off, by the way. After news of the team's results hit the headlines, X-Tek's BladeRunner technology received a burst of interest from companies who wanted to use it to check aeroplane and spacecraft parts. Hadland found a buyer he trusted and sold the company in December 2007. He has stayed on as a consultant, meaning that he can now spend his time doing what he loves best – designing new machines – although he's not sure he can ever beat working on the Antikythera mechanism, which he describes as 'the crowning glory of my career'.

Meanwhile, Michael Wright has added some new features to his model, including a dial showing the day of the month, and he is working on further publications on the details of the gearwork. He also plans to make a second reconstruction out of bronze, which he hopes to present to the Athens National Museum. He remains convinced that the main purpose of the mechanism was to show the movements of the planets. Fragment D contains the only wheel (or possibly two identical wheels, one above the other) that doesn't fit anywhere in the latest reconstruction of the mechanism. Wright believes that it may be part of the lost planetary gearing on the front of the device. The wheel has 63 teeth, which would fit within an epicyclic train for calculating the motion of Mercury.

Wright still thinks that the mechanism could have been put together from the pieces from two or three other devices – partly because of the way that the wood in the case is jointed. And he has seen something intriguing, relating to the subsidiary dial on the eclipse spiral. Tony Freeth's team saw that this dial was divided into three, with the numbers 8 and 16 inscribed in two of the sectors, to show the number of hours that had to be added to the predicted eclipse times depicted on the main spiral. As far as their X-ray images show, the third sector is blank. But Wright, unlike any of the later researchers, was able to examine the Antikythera fragments by eye. He is convinced that – hidden under an overhang of limestone – he saw the edge of a character in the third sector. If he's right, it would be the first known Greek use of a symbol for zero (that honour currently goes to Ptolemy, in the second century AD).

All of the researchers hope that in the future new pieces of evidence will be discovered. It's possible that more missing fragments of the Antikythera mechanism might be found at the Athens National Museum, lying unrecognised in the stores since they were brought up from the sunken ship in 1901. The wreck, too, might yet yield more secrets. The hull of the ship and some of its contents still appear to be intact and buried beneath the mud, so modern divers equipped with the right tools could uncover more treasures and – who knows – even another astronomical computer. There's also the hope that more ancient wrecks will be discovered in the Mediterranean, ideally in deeper water out of the reach of looters, and that as news of the Antikythera mechanism spreads, fragments from geared devices throughout history might even resurface in basements or museum stores as owners recognise their potential significance.

But the richest source of new information may turn out to be old Islamic manuscripts. Work to interpret these is in its infancy and there are thousands of manuscripts that have never been catalogued, let alone read: few people combine the language skills to translate these documents with the technical expertise needed to understand their contents. An Arabic translation of Archimedes' lost treatise on sphere-making, for instance, might solve the mystery of the origin of this ancient technology once and for all.

Acknowledgements

I WOULD LIKE to thank my agent, Peter Tallack at The Science Factory, whose idea it was that I write this book and without whose enthusiasm and support it probably wouldn't have happened. Thanks to my editors, Jason Arthur at William Heinemann and Bob Pigeon at Da Capo Press, for invaluable editorial comments and for believing in the book enough to publish it, and to Laurie Ip Fung Chun for helping with pictures, references and a hundred other things. And thanks to Oliver Morton at *Nature*, where this all started, for first sending me to Athens to write about a strange contraption called the Antikythera mechanism.

I am greatly indebted to Michael Wright, a gentleman who fielded my never-ending questions with honesty and grace, and to his wife Anne. As well as sharing his personal story with me, Michael shared his ideas about how the Antikythera mechanism worked, where it came from and how widespread such devices might have been, including the arguments relating to the size of the mechanism and the statistics of such finds described in chapter 10. He provided some photographs, much of the information used in the diagrams, and many helpful corrections and comments on the manuscript itself.

Thanks to Tony Freeth and his colleagues – Mike Edmunds, Yanis Bitsakis, Xenophon Moussas and Agamemnon Tselikas – who were charming company and of great help when I first researched the Antikythera mechanism. Once I started writing this book, they felt unable to speak to me any further about their work or to be involved in any way. I hope I have done justice to their roles.

I am grateful to Roger Hadland for sharing the story of his company, X-Tek, and its role in the Antikythera project, for his helpful comments on chapter 8, and for providing photographs of BladeRunner's trip to Athens.

Many others were kind enough to share their knowledge and ideas with me, including Jonathan Adams of the University of Southampton, Phaedon Antonopoulos of the Hellenic Institute of Marine Archaeology, Alexander Apostolides of the London School of Economics, Jane Biers formerly of the University of Missouri, Mary Ellen Bowden of the Chemical Heritage Foundation, Alexis Catsambis of Texas A&M University, Francois Charette of the University of Frankfurt, Serafina Cuomo of Imperial College London, JV Field of Birkbeck University, Eugene Garfield of the Institute for Scientific Information, Owen Gingerich of the Harvard-Smithsonian Center for Astrophysics, Bert Hall of the University of Toronto, Robert Hannah of the University of Otago, David King formerly of the University of Frankfurt, Stephen Johnston of the Museum for the History of Science, Alexander Jones of the Institute for the Study of the Ancient World, Eleni Magkou of the National Archaeological Museum, Stefanie Maison, Tom Malzbender of Hewlett Packard, Basim Musallam of the University of Cambridge, Emmanuel Poulle of the École Nationale des Chartes, Andrew Ramsey of X-Tek, David Sedley of the University of Cambridge, John Steele of the University of Durham, Peter Stewart of the Courtauld Institute of Art, Doron Swade formerly of the Science Museum, Sharon Thibodeau of the US National Archives and Records Administration, Natalie Vogeikoff-Brogan of the American School of Classical Studies at Athens, Faith Warn, author of *Bitter Sea*, and Mairi Zafeiropoulou of the National Archaeological Museum. I am especially grateful also for the help of Anthony Michaelis and Arthur C. Clarke; sadly, neither of them lived to see the finished book.

The staff of the National Archaeological Museum in Athens, the British Library, the V&A Library, the Senate House Library, Sam Fogg's in London and the Adler Planetarium in Chicago were extremely helpful while I was researching this book. Anne Bromley, Joy Elliott, Brendan Foley of Woods Hole Oceanographic Institution, Iain Macquarrie of www.divingheritage.com, Panos Travlos of Travlos Publishers and Doron Swade all provided photographs for the book or helped in my search. Special thanks to Paul Cartledge for making several corrections to the text.

Thanks to my parents, Jim and Diana Marchant, especially for putting up with me writing in their house all through Christmas. Finally thank you to Ian Sample, for everything.

Picture Credits

GRATEFUL ACKNOWLEDGEMENT IS made to the following for permission to reprint photographs:

Captain Dimitrios Kontos and his crew of sponge divers, from 'Das Athener Nationalmuseum' by J.N. Svoronos (Athens, 1908)

A Greek sponge diver from the early 20th century, reproduced with permission of www.divingheritage.com

Bronze portrait head from a statue of a philosopher © B. Foley

Bronze statue nicknamed the Antikythera Youth © Jo Marchant

Marble statue of a crouching boy © Jo Marchant

Fragment C of the Antikythera mechanism © Jo Marchant

Fragment B of the Antikythera mechanism © Jo Marchant

Fragment A of the Antikythera mechanism © Jo Marchant

Derek de Solla Price © Malcolm S. Kirk/Artsmarket

The Tower of the Winds © Jo Marchant

Allan Bromley © Steven Siewert/Fairfaxphotos

Michael Wright © Anne Wright

Main frame of Wright's model © Michael Wright

Part of Wright's model © Michael Wright

Islamic geared astrolabe reproduced by permission of the Museum of the History of Science, University of Oxford

The Nastulus Manuscript showing the 'Box of the Moon' © Sam Fogg, London

Alan Crawley © Roger Hadland

Pandelis Feleris © Roger Hadland

Dr Eleni Magkou (National Archaeological Museum) and Professor Xenophon Moussas (National University of Athens) © Roger Hadland

Surface of the Antikythera fragments superimposed onto X-ray © Nature Publishing Group

Every effort has been made to contact all copyright holders. If notified, the publisher will be pleased to rectify any errors or omissions at the earliest opportunity.

Sources and Further Reading

Chapter 1

Books:

- Cousteau, Jacques, *The Silent World* (Elm Tree Books, 1988)
- Homer (translated by Walter Shewring), *The Odyssey* (Oxford paperbacks, New edition, 1998)
- Svoronos, J.N., *Das Athener Nationalmuseum* (Athens: Beck and Barth, 1908)
- Warn, Faith, *Bitter Sea: The Real Story of Greek Sponge Diving* (Guardian Angel, 2000)

Papers:

- Catsambis, Alexis, 'Before Antikythera: The first underwater archaeological survey in Greece', in *The International Journal of Nautical Archaeology*, vol. 35, issue 1, pp. 104–7 (2006)
- Cornelius Bol, Peter, 'Die Skulpturen des Schiffsfundes von Antikythera', in *American Journal of Archaeology*, vol 77, no. 4, pp. 451–453 (Oct 1973)
- Frost, K.T., 'The statues from Cerigotto,' in *The Journal of Hellenic Studies*, vol. 23, pp. 217–236 (1903)
- Kavvodias, P., 'The recent finds off Cythera', in *The Journal of Hellenic Studies*, vol. 21, pp. 205–208 (1901)

- Karo, George, 'Art salvaged from the sea', in *Archaeology*, vol. 1, pp. 179–185 (1948)
- 'The findings of the wreckage of Antikythera', in *Report of the Archaeological Society of Athens,* in Greek (15 Feb 1902)

Chapter 2

Books:

- Diels, Hermann, *Antike Teknik* (Leipzig and Berlin, 1920)
- Gunther, Robert, *Astrolabes of the World* (Oxford, 1932)
- Schlachter, Alois, *Der Globus*, (Leipzig, 1927)
- Svoranos, J.N., *Das Athener Nationalmuseum* (Athens: Beck and Barth, 1908)
- Zinner, Ernst, *Geschichte der Sternkunde* (Berlin, 1931)

Papers:

- Luce, Stephen B., 'Albert Rehm', in *American Journal of Archaeology*, vol. 54, no. 3, p. 254 (Jul–Sep 1950)
- Neugebauer, Otto, 'The early history of the Astrolabe. Studies in ancient Astronomy IX', in *Isis*, vol. 40, no. 3, pp. 240–256 (Aug 1949)
- Théofanidis, Jean, 'Sur l'instrument en cuivre dont des fragments se trouvent au Musée Archéologique d'Athènes et qui fut retiré du fond de la mer d'Anticythère en 1902', in *Praktika tes Akademias Athenon*, vol. 9, pp. 140–154 (Athens, 1934)

Chapter 3

Books:

- Berthold, Richard M., *Rhodes in the Hellenistic Age* (Cornell University Press, 1984)

- Dumas, Frédéric, *30 Centuries Under the Sea* (Crown, 1976)
- Lane Fox, Robin, *The Classical World: An Epic History of Greece and Rome* (Penguin Books, 2006)
- Plutarch, *The Parallel Lives* (Loeb Classical Library Edition, 1916)
- Throckmorton, Peter (ed.), *The Sea Remembers: Shipwrecks and Archaeology* (Chancellor Press, 1987)

Papers:

- Basch, Lucien, 'Ancient wrecks and the archaeology of ships', in *The International Journal of Nautical Archaeology and Underwater Exploration*, vol. 1, pp. 1–58 (1972)
- Bass, George F., Peter Throckmorton et al., 'Cape Gelidonya: A Bronze Age Shipwreck', in *Transactions of the American Philosophical Society*, New Ser., vol. 57, no. 8, pp. 1–177 (1967)
- Davidson Weinberg, Gladys, Virginia R. Grace, G. Roger Edwards, Henry S. Robinson, Peter Throckmorton, Elizabeth K. Ralph, 'The Antikythera Shipwreck reconsidered', in *Transactions of the American Philosophical Society*, New Ser., vol. 55, no. 3 (1965)
- Ermoupolites, Mr, 'The amphorae of the Antikythera wreck', in *Naftiki Hellas*, in Greek (Aug 1950)
- Gibbins, David and Jonathan Adams, 'Shipwrecks and maritime archaeology', in *World Archaeology*, vol. 32, issue 3, pp. 279–291 (2001)
- Koehler, Carolyn G., 'Virginia Randolph Grace, 1901–1994', in *American Journal of Archaeology*, vol. 100, pp. 153–155 (1996)
- Yalouris, N., 'The shipwreck of Antikythera: New evidence of its date after supplementary investigation', in Jean Paul Descoeudres, (ed.) *Eumousia: Ceramic and Iconographic Studies in Honour of Alexander Cambitoglou*, pp. 135–136 (Sydney: Meditarch, 1990)

Chapter 4

Books:

- McAleer, Neil, *Odyssey: The Authorised Biography of Arthur C. Clarke* (Victor Gollancz, 1992)
- Price, Derek J. de Solla, *Little Science, Big Science* (Columbia University Press, 1963)
- Price, Derek J. de Solla, *Little Science, Big Science ... and Beyond*, Foreword by Robert K. Merton and Eugene Garfield (Columbia University Press, 1986)
- Price, D.J.D., *The Equatorie of the Planetis* (Cambridge University Press, 1955)
- Welfare, Simon and John Fairley, *Arthur C. Clarke's Mysterious World* (Fontana, 1980)

Papers:

- Drachmann, A. G., 'The plane astrolabe and the anaphoric clock', in *Centaurus*, vol. 3, p. 183 (1954)
- Garfield, Eugene, 'A tribute to Derek John de Solla Price: A bold, iconoclastic historian of science', in *Essays of an Information Scientist*, vol. 7, pp. 213–217 (1984)
- Garfield, Eugene, 'Derek Price and the practical world of scientometrics', in *Science, Technology and Human Values 13 (3/4)*, pp. 349–50 (1988)
- Garfield, Eugene, 'In memoriam', in *Essays of an Information Scientist*, vol. 6, p. 645 (1983)
- Landels, J. G., 'Water clocks and time measurement in classical antiquity', in *Endeavour*, New Series vol. 3, no. 1 (1979)
- Morris, Robert L., 'Derek de Solla Price and the Antikythera mechanism: An appreciation', in *IEEE Micro*, pp. 15–21 (Feb 1984)

- Needham, J., W. Ling and D.J.D. Price, 'Chinese astronomical clockwork', in *Nature*, vol. 177, pp. 600–2 (1956)
- Noble, Joseph V. and Derek J. de Solla Price, 'The water clock in the Tower of the Winds', in *American Journal of Archaeology*, vol. 72 no. 4, pp. 345–355 (Oct 1968)
- Price, Derek, 'Editorial statements', in *Scientometrics*, vol. 1, pp. 3–8 (Sept 1978)
- Price, Derek J. de Solla, 'Clockwork before the clock', in *Horological Journal* (5 Oct 1955)
- Price, Derek J. de Solla, 'An Ancient Greek computer', in *Scientific American*, pp. 60–67 (1959)
- Price, D.J.D., 'The equatorium of the planetis', in *Bull. Brit. Soc. Hist. Sci.*, vol. 1, pp. 223–6 (1953)
- Price, D.J.D., 'Networks of scientific papers', in *Science*, vol. 149, pp. 510–5 (1965)
- Price, Derek J., 'The prehistory of the clock', in *Discovery,* pp. 153–157 (April 1956)
- Price, Derek J. de Solla, 'The tower of the winds: Piecing together an ancient puzzle', in *National Geographic*, pp. 586–596 (April 1967)

Chapter 5

Books:

- Däniken, Erich von, *Chariots of the Gods? Memories of the Future Unsolved Mysteries of the Past* (G.P. Putnam's Sons, 1969)
- Däniken, Erich von, *Odyssey of the Gods: The Alien History of Ancient Greece* (Vega, 2002)
- Feynman, Richard, *What Do You Care What Other People Think?* (Unwin Hyman, 1989)

- Neugebauer, Otto, *A History of Ancient Mathematical Astronomy* (Springer, 1975)

Papers:

- Beaver, Donald deB., 'Eloge: Derek John de Solla Price', in *Isis*, vol. 76, issue 3, pp. 371–374 (1985)
- 'Interview: Derek de Solla Price', in *Omni*, pp. 89–102, 136 (1982)
- MacKay, Alan, 'Derek John de Solla Price: An appreciation', in *Social Studies of Science*, vol. 14, pp. 315–20 (1984)
- Miller, F.J., E.V. Sayre and B. Keisch, 'Isotopic methods of examination and authentication in art and archaeology', in *Oak Ridge National Laboratory* IIC-21 (Oak Ridge, Oct 1970)
- Price, Derek J. de Solla, 'Gears from the Greeks. The Antikythera Mechanism: A calendar computer from ca. 80 BC', in *Transactions of the American Philosophical Society*, New Ser., vol. 64, no. 7, pp. 1–70 (1974)
- Shapley, Deborah, 'Nuclear weapons history: Japan's wartime bomb projects revealed', in *Science*, vol. 199, pp. 153–157 (1978)
- Shizume, Eri Yagi and Derek J. de Solla Price, ' Japanese bomb', in *Bulletin of the Atomic Scientists*, p. 29 (Nov 1962)
- 'The Leonardo da Vinci Medal', in *Technology and Culture*, pp. 471–478 (1976)

Chapter 6

Books:

- Asprey, William (ed.), *Computing Before Computers* (Iowa State University Press, 1990)
- Swade, Doron, *The Cogwheel Brain: Charles Babbage and the Quest to Build the First Computer* (Abacus, 2001)

- Wright, M.T., J.V. Field and D.R. Hill, *Byzantine and Arabic Mathematical Gearing* (The Science Museum, 1985)

Papers:

- Bromley, Allan G., 'Notes on the Antikythera mechanism', in *Centaurus*, vol. 29, pp. 5–27 (1986)
- Bromley, Allan G., 'The Antikythera Mechanism: A reconstruction', in *Horological Journal*, p. 28–31 (July 1990)
- Bromley, Allan G., 'Observations of the Antikythera mechanism', in *Antiquarian Horology*, pp. 641–652 (Summer 1990)
- Bromley, Allan G., 'The Antikythera mechanism', in *Horological Journal*, p. 412–415 (June 1999)
- Cherfas, Jeremy, 'Seeking the soul of an old machine', in *Science*, New Series, vol. 252, no. 5011, pp. 1370–1371 (7 June 1991)
- Edmunds, Mike and Philip Morgan, 'The Antikythera mechanism: Still a mystery of Greek astronomy?', in *Astronomy and Geophysics*, vol. 41, pp. 6.10–6.17 (Dec 2000)
- Field, J.V. and M. T. Wright, 'Gears from the Byzantines: A portable sundial with calendrical gearing', in *Annals of Science*, vol. 42, pp. 87–138 (1985)
- Maddison, Francis, 'Byzantine calendrical gearing', in *Nature*, vol. 314, pp. 316–317 (1985)
- Wright, M.T., A.G. Bromley and H. Magou, 'Simple X-ray tomography and the Antikythera mechanism', in *PACT (Journal of the European Study Group of Physical, Chemical, Biological and Mathematical Techniques Applied to Archaeology)*, pp. 45, 531–543 (1995)

Chapter 7

Books:

- Evans, James, *The History and Practice of Ancient Astronomy* (Oxford University Press, 1998)

Papers:

- Wright, M.T. and A.G. Bromley, 'Current work on the Antikythera mechanism', in *Proc. Conf. on Ancient Greek Technology*, pp. 19–25 (Greece, Thessaloniki, Sept 1997)
- Wright, M.T., 'A planetarium display for the Antikythera mechanism', in *Horological Journal,* pp. 144, 169–173 and 193 (2002)
- Wright, M.T. 'Epicyclic gearing and the Antikythera mechanism, Part I', in *Antiquarian Horology*, vol. 27, pp. 270–279 (2003)
- Wright, M.T., 'In the steps of the master mechanic', in *Proc. Conf. on Ancient Greece and the Modern World*, pp. 86–97 (Greece: University of Patras, 2003)
- Wright, M.T. and A.G. Bromley, 'Towards a new reconstruction of the Antikythera mechanism', in *Extraordinary Machines and Structures in Antiquity* (ed. S.A. Paipetis), pp. 81–94 (Patras: Peri Technon, 2003)
- Wright, M.T., 'The scholar, the mechanic and the Antikythera mechanism', in *Bulletin of the Scientific Instrument Society*, vol. 80, pp. 4–11 (2004)
- Wright, M.T., 'Counting months and years: The upper back dial of the Antikythera mechanism', in *Bulletin of the Scientific Instrument Society*, vol. 87, pp. 8–13 (2005)
- Wright, M.T., 'Epicyclic gearing and the Antikythera mechanism, Part 2', in *Antiquarian Horology*, vol. 29, pp. 51–63 (2005)

- Wright, M.T., 'The Antikythera mechanism: A new gearing scheme', in *Bulletin of the Scientific Instrument Society*, vol. 85, pp. 2–7 (2005)
- Wright, M.T., 'The Antikythera mechanism and the early history of the moon phase display', in *Antiquarian Horology*, vol. 29, pp. 319–329 (2006)
- Wright, M.T. ,'Understanding the Antikythera mechanism', in *Proc. 2nd Int. Conf. on Ancient Greek Technology*, pp. 49–60 (Athens: Technical Chamber of Greece, 2006)
- Wright, M.T. 'The Antikythera mechanism reconsidered', in *Interdisciplinary Science Review*, vol. 32, no. 1, pp. 27–43 (2007)

Chapter 8

Papers:

- Brooks, Michael, 'Tricks of the light', in *New Scientist*, pp. 38–41 (7 April, 2001)
- Edmunds, Mike, 'The elementary universe', in *Astronomy and Geophysics*, vol. 46 pp. 4.12–4.17 (Aug 2005)
- Edmunds, Mike, 'Landscapes, circles and Antikythera: The birth of the mechanical universe', in *Mediterranean Archaeology and Archaeometry*, Special Issue, vol. 6, no. 3, pp. 87–92 (2006)
- Freeth, Tony, 'The Antikythera mechanism. I. Challenging the classic research', in *Mediterranean Archaeology and Archaeometry*, vol. 2, no. 1, pp. 21–35 (2002)
- Ramsey, Andrew T., 'The latest techniques reveal the earliest technology – A new inspection of the Antikythera mechanism', in *International Symposium on Digital Industrial Radiology and Computed Tomography*, pp. 25–27 (France, Lyon, June 2007)

- Seabrook, John, 'Fragmentary knowledge: Was the Antikythera mechanism the world's first computer?', in *The New Yorker* (14 May, 2007)
- Solomos, N.H. (ed), 'The Antikythera Mechanism – Real Progress Through Greek/UK/US Research. M.G. Edmunds for the Antikythera Research Project. Recent Advances in Astronomy and Astrophysics', in *7th International Conference of the Hellenic Astronomical Society*, pp. 913–918 (2006)

Chapter 9

Books:

- Netz, Reviel and William Noel, *The Archimedes Codex: Revealing the Blueprint of Modern Science* (Phoenix, 2007)
- Steele, John, *Observations and Predictions of Eclipse Times by Early Astronomers* (Kluwer Academic Publishers, 2000)

Papers:

- Charette, Francois, 'High-tech from Ancient Greece', in *Nature*, vol. 444, pp. 551–552 (2006)
- Freeth, T., Y. Bitsakis, X. Moussas, J.H. Seiradakis, A. Tselikas, H. Mangou, M. Zafeiropoulou, R. Hadland, D. Bate, A. Ramsey, M. Allen, A. Crawley, P. Hockley, T. Malzbender, D. Gelb, W. Ambrisco, M.G. Edmunds, 'Decoding the Ancient Greek astronomical calculator known as the Antikythera mechanism', in *Nature*, vol. 444, pp. 587–591 (2006)
- Marchant, Jo, 'In search of lost time', in *Nature* vol. 444, pp. 534–538 (2006)
- Steele, John, 'Eclipse prediction in Mesopotamia', in *Archive for History of Exact Sciences*, vol. 54, no. 5, pp. 421–454 (Feb 2000)

- Steele, John, 'Ptolemy, Babylon and the rotation of the earth', in *Astronomy and Geophysics*, vol. 46 pp. 5.11–5.15 (2005)

Chapter 10

Books:

- Cicero, *The Nature of the Gods* (Oxford University Press, 1998)
- Evans, James, *The History and Practice of Ancient Astronomy* (Oxford University Press, 1998)
- Hill, Donald, *Islamic Science and Engineering* (Edinburgh University Press, 1993)
- Kidd, Ian Gray, *Posidonius: Volume III The Translation of the Fragments* (Cambridge University Press, 2004)
- Linssen, Marc J.H., *The Cults of Uruk and Babylon: The Temple Ritual Texts as Evidence for Hellenistic Cult Practice* (Brill-Styx, 2004)
- Pliny, *Natural History* (Penguin, 1991)
- Rosheim, Mark E., *Robot Evolution: The Development of Anthrobotics* (Wiley-Interscience, 1994)
- Ruggles, Clive, *Astronomy in Prehistoric Britain and Northern Ireland* (Yale University Press, 1999)
- Russo, Lucio, *The Forgotten Revolution: How Science Was Born in 300 BC and Why it Had to Be Reborn* (Springer, 2003)
- Toomer, G.J. (translated), *Ptolemy's Almagest* (Princeton University Press, 1998)

Papers:

- Freeth, Tony, Alexander Jones, John Steele and Yanis Bitsakis, 'Calendars with Olympiad display and eclipse prediction on the Antikythera mechanism', in *Nature*, vol. 454, pp. 614–617 (2008)

- Gingerich, Owen, 'Islamic astronomy', in *Scientific American*, vol. 254, p. 74 (April 1986)
- Jones, Alexander, 'The adaptation of Babylonian methods in Greek numerical astronomy', in *Isis*, vol. 82, pp. 441–453 (1991)
- Jones, Alexander, 'The astronomical inscription from Keskintos, Rhodes', in *Mediterranean Archaeology and Archaeometry*, Special Issue, vol. 6, no. 3, pp. 213–220 (2006)
- Jones, Alexander, 'The Keskintos astronomical inscription text and interpretations', in *SCIAMVS*, vol. 7, pp. 3–41 (2006)
- Keyser, Paul, 'A new look at Heron's "Steam Engine"', in *Archive for History of Exact Sciences*, vol. 44, pp. 107–124 (1992)
- Price, Derek J. de Solla 'Automata and the origins of mechanism and mechanistic philosophy', in *Technology and Culture*, vol. 5, no. 1, pp. 9–23 (Winter 1964)
- Toomer, G.J., 'Hipparchus', in *Dictionary of Scientific Biography*, ed. Charles Gillespie, vol. XV, pp. 207–224 (New York: Scribners, 1970–1980)
- Toomer, G.J., 'Hipparchus and Babylonian astronomy', in *A Scientific Humanist: Studies in Memory of Abraham Sachs*, pp. 353–362 (Philadelphia: Occasional Publications of the Samuel Noah Kramer Fund, 9)
- Tybjerg, Karin, 'Wonder-making and philosophical wonder in Hero of Alexandria', in *Stud. Hist. Phil. Sci.*, vol. 34, pp. 443–446 (2003)
- Waerden, Bartel van der, 'Mathematics and astronomy in Mesopotamia', in *Dictionary of Scientific Biography* (ed. in Chief Charles Gillespie), vol. XV, Supplement I, pp. 667–680 (1978)

Index